Policy Issues and Challenges for Interagency Space Systems Acquisition

Dana J. Johnson, Gregory H. Hilgenberg, Liam P. Sarsfield

Prepared for the National Reconnaissance Office

RAND

National Security Research Division

The research described in this report was conducted for the Deputy Director, Air Force-NRO Integration Planning Group (ANIPG), within the Acquisition and Technology Policy Center of RAND's National Security Research Division under Contract NR0-000-98-D-2628.

ISBN: 0-8330-3011-6

Published 2001 by RAND
1700 Main Street, P.O. Box 2138, Santa Monica, CA 90407-2138
1200 South Hayes Street, Arlington, VA 22202-5050
201 North Craig Street, Suite 102, Pittsburgh, PA 15213-1516
RAND URL: http://www.rand.org/
To order RAND documents or to obtain additional information, contact Distribution Services: Telephone: (310) 451-7002; Fax: (310) 451-6915; Internet: order@rand.org

Preface

In the recent past, there have been increasing pressures for agencies within the national security space community to conduct joint or interagency programs, with the goal being to take advantage of potentially shared objectives and mission synergies. Expectations are that such joint or interagency programs will result in improvements in efficiency and effectiveness and the elimination of unnecessary redundancies among programs. However, the process of undertaking such efforts needs to recognize a number of policy issues and challenges that will influence how these programs are executed. This study seeks to illuminate these policy issues and challenges, particularly in how they will influence multi-mission space system concepts and programs conducted jointly by the National Reconnaissance Office (NRO) and the Air Force.

The research for this report was conducted for the Deputy Director, Air Force-NRO Integration Planning Group (ANIPG), and was performed in the Acquisition and Technology Policy Center of RAND's National Security Research Division.

Contents

Figures

Tables

Summary

Overview and Approach

A number of trends and recent events are providing the motivation behind consideration of interagency programs by agencies of the U.S. Government. These trends include the increasing number of contingencies that U.S. military forces and the intelligence community are supporting, overall change in focus and increase in long term intelligence requirements worldwide, the growth in commercial space activities, and increased congressional and Administration scrutiny of space and intelligence programs. Since 1996, the Air Force and the NRO have been addressing air and space integration, that is, the efficient and effective operational delivery of aerospace capability to the user, whether a warfighter or a national decisionmaker. Consequently, the Air Force and the NRO have been undertaking the identification of possible concepts and programs for cooperation and collaboration in aerospace integration between the two agencies.

Opportunities for conducting coordinated or integrated activities, such as developing concepts of operations (CONOPs), determining required capabilities, or developing integrated acquisition programs can occur at different points in time. This process is iterative, as shown in Figure S.1 below, for the stakeholders and users provide important feedback on the success or failure of a particular program. The interagency program office (IPO) is another means in this process by which to execute cooperative activities between organizations sharing common interests and goals. IPOs may be governed by laws, regulations, and policies, and certain government organizations like the Department of Defense (DoD) possess structured, rigorous processes for joint programs. The complexity for IPOs arises when conducting cooperative or collaborative activities among agencies with very different acquisition and budgetary processes, organizational structures and cultures, and stakeholder/user bases. While there may be no one "best" way to conduct IPOs, insights can be gained by analyzing prior and ongoing joint or interagency programs for application by the NRO and the Air Force.

What are the key decisions in pursuing and implementing an integrated program concept and how should the NRO and the Air Force make those decisions?

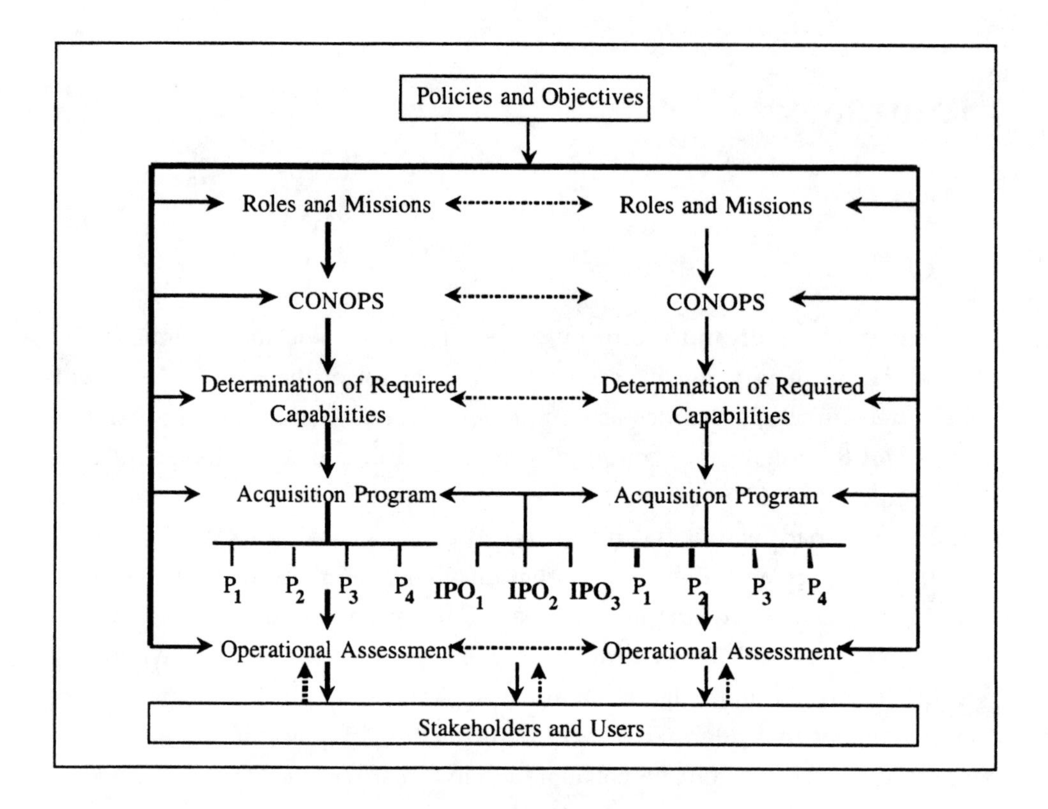

Figure S.1 Opportunities for Integration

Furthermore, what criteria can be identified that measure progress towards successful integration, whether across institutions, in military and intelligence operations, or within an individual program? High-level functional decisions about what activities or programs to pursue via an IPO will drive organizational decisions about the development and implementation of IPO organizational structures. These decisions lead to cascading decisions about the management process guiding the IPO, the staff management process, and approaches to satisfying and maintaining external stakeholder support for the IPO program.

This report seeks to provide an analytical foundation for the implementation of collaborative concepts between the Air Force and the NRO by an examination of the mechanism of an IPO. Using the approach of a "SWOT chart," a business school analytical approach that examines strengths, weaknesses, opportunities, and threats to particular programs, the report identifies important considerations for undertaking interagency programs and assesses five approaches to the interagency program office concept. (A sixth approach was noted but not addressed in detail.) The considerations are divided into seven elements that are addressed individually and then used to compare and contrast the IPO approaches. These elements consist of:

- *Acquisition complexity*: denotes the degree of difficulty involved in acquiring a particular program or capability; includes efforts to ensure that the program satisfies existing policy and objectives guidance and "front end" incentives to form the IPO
- *Program management*: refers to the organization, structure, and approach taken within a program to accomplish objectives
- *Program control*: defined as the ability to monitor and influence the operation of a program by the responsible individual, i.e., a program manager; this is also called "span of control"
- *Requirements management*: involves the adjudication, coordination, and implementation of a common requirements process for the program
- *Funding stability*: defined as the process of maintaining funding support among the organizational partners over the lifetime of the program
- *Customer responsiveness*: denotes the program's relationship to its user base, e.g., how supportive are the users and stakeholders for the program
- *Cultural alignment*: the interaction of and implications for the program of the diverse organizational cultures inherited from parent or partner organizations
- *Staffing*: includes the staffing process of the program and the ability to attract qualified personnel to work in the program

Generic questions for each element are raised as illustrations of the deliberative process necessary first, to determine whether circumstances are right for an interagency program, and secondly, as potential metrics for evaluating a chosen IPO approach. For example, under "acquisition complexity," the types of questions raised include:

- How important is the proposed program to the missions of the parent organizations?
- What policies provide rationale for the IPO?
- What mission requirements does the IPO satisfy?
- Have incentives to form the IPO been identified?

For "program management," the types of questions include those that address the development and implementation of a Memorandum of Understanding (MOU) or a Memorandum of Agreement (MOA), the establishment of cost-sharing arrangements between the parent or participating organizations, and the organizational structure of the IPO. The other elements have similar kinds of questions. Subsequently, these questions are used to broadly evaluate the alternative IPO approaches and to derive insights for application in the implementation of an IPO by the Air Force and the NRO.

Five alternative approaches to interagency programs are examined in the report. These approaches include *Executing Agent, System Integrator, Independent Agent, Confederation,* and *Joint Program Office,* and are defined in Table S.1 below. The sixth approach, *Commercial Prime,* was identified but not analyzed in depth; nevertheless, we believe this approach deserves further consideration as a possible IPO alternative.

Table S.1 Definition of Interagency Program Office Alternative Approaches

Type	*Description*
Executing Agent	Agency designated lead for technology demonstration, development, acquisition, and/or operation of program for common or multi-user needs
System Integrator	Joint venture partners build system elements with lead organization operating as integrator
Independent Agent	Creation of new, independent, functionally focused entity to acquire, execute, operate program
Commercial Prime	Government partner using commercial development vehicle
Confederation	Multiple entities form acquisition "alliance" to accomplish limited, albeit challenging, objectives
Joint Program Office	Single, integrated program independent of, but responsive to, parent organizations

Table S.2 further defines each alternative along the lines of different constructs, such as organizational structure (i.e., whether there is a single participant or multiple participants), policy and regulatory approaches (whether traditional or streamlined), and missions, i.e., whether the program is R&D or a demonstration, a small system buy (e.g., very limited numbers of spacecraft), a multiple system buy (e.g., "block buy" of large orders of spacecraft), or a data buy (e.g., buying imagery from a variety of government and commercial sources). Examples from

Table S.2 Comparison of Alternative Acquisition Approaches and Illustrative Examples

Alternative Acquisition Approaches	**Alternative Constructs and Examples**							
	Organizational Structure		**Policies and Regulations**		**Mission**			
	Single Participant	Multiple Participants	Traditional	Streamlined	R&D/ Demonstration	Small System Buy	Large System Buy	Data Buy
Executing Agent	ACTDs	ACTDs		ACTDs	ACTDs			
System Integrator		ISSP	ISSP			ISSP		
Independent Agent	NIMA		NIMA	NIMA			NIMA	NIMA
Commercial Prime	Multiple Examples							
Confederation		NSLRS-D	NSLRS-D					NSLRS-D
Joint Program Office		GPS N-POESS Discoverer II Arsenal Ship	GPS N-POESS	Discoverer II Arsenal Ship	Discoverer II Arsenal Ship	N-POESS	GPS	

Legend: ACTD = Advanced Concept Technology Demonstration; ISSP = International Space Station Program; NIMA = National Imagery and Mapping Agency; NSLRSD = National Satellite Land Remote Sensing Data Archive; GPS = Global Positioning System; NPOESS = National Polar-orbiting Environmental Satellite System.

prior or ongoing programs are to illustrate specific points in each of the approaches. The "real world" examples tend to fall in certain alternatives, largely because some of the alternatives as we have defined them would likely be very difficult to implement. Consequently, we devote more attention to those at the expense of some of the others. Nevertheless, all offer interesting analytical aspects worthy of examination.

Insights and Observations

Numerous insights and observations stand out as being important or even essential to IPO success, and are consistent with both the SWOT-related questions and as evidenced by the case studies and background research conducted for this study. Any consideration of a potential integration concept involving the use of an IPO should weave the elements together into a coherent program strategy that makes sense for the particular concept in mind. Again, while this may be a truism for most acquisition programs, the multi-agency nature of an IPO will complicate every element of the program strategy, but it should not be considered an insurmountable problem.

Acquisition Complexity

The importance and criticality of support from the leadership of each participating organization cannot be underestimated. Each partner agency's management must understand and accept the organizational agreements negotiated among the partners, for their support is integral to program success. When it comes to determining specific responsibilities, roles, and responsibilities for program activities at the parent organization level, high level organizational leadership support is also necessary for ensuring appropriate cooperation at lower levels within the parent organizations. This becomes important particularly during budgetary reviews for ensuring adequate funding stability and continuity.

Program goals and objectives should be consistent with higher-level policy guidance, including national space policy and national security policy, defense planning guidance, relevant intelligence policies, and legal and regulatory agreements such as treaties, where appropriate. Of particular interest will be policies or regulations that guide or bind one partner organization but not the other. Security and interoperability considerations should be factored into the planning throughout, and potential concerns resolved. Linked to security considerations are jointly agreed-to mechanisms for maintaining IPO information infrastructure assurance against common threats. A survey of related or similar activities underway in other organizations should be conducted early on to identify unique applications and potential opportunities for further collaboration. For example, identifying programs or technology demonstration efforts underway in civil space agencies such as NASA may lead to useful collaboration in solving particularly difficult or challenging technical problems. This has the added benefit of further expanding program support from other agencies.

Program Management

Following agreement on common program goals and agency interests, addressing program management aspects, specifically the MOU/MOA, is crucial to establishing the scope and organizational structure of the IPO. The MOU/MOA needs to be sufficiently robust to ensure parent agency support and to assign specific roles and responsibilities among participating agencies, yet flexible enough to respond to potential changes in policy and planning guidance, the threat, or other high-level factors. Also key is the development of an overall program strategy that includes implementation and funding strategies running the lifetime of the program, including termination, is consistent with national policy and guidance, agency goals and objectives, and regulations, and is executable at critical program milestones. The program funding strategy should include cost sharing arrangements among the parent agencies, and include possible penalties for withdrawal from the agreed-to arrangement (again, to encourage and facilitate overall funding stability).

Program planning and management also need to recognize and account for the potential challenges posed by differing parent agency planning and budgetary cycles and the increased burden in time and manpower placed on IPO leadership to deal with maintaining funding stability and stakeholder support. Furthermore, requirements adjudication will be a key issue to which the IPO program leadership needs to devote significant attention and resources.

Program Control

Program control addresses the flow of information and communication through the integrated organizational chain of command, and is influenced by organizational structure and the program management's "span of control." The approach to program control also can influence or hinder the collocation of authority and responsibility to encourage accountability within the program.

The program strategy discussed earlier needs to ensure that the program manager and his/her senior team have unimpeded access to the information they need to execute the program successfully. Complicating program control will be changes in laws and regulatory policies regarding privacy and confidentiality requirements, liability, and national security requirements, and the overall growth in information technologies and access to information via the Internet.

Organizationally, a strong, decisive executive council is needed to keep the program on track and to engage at senior levels of the parent agencies and elsewhere as appropriate to deflect potential problems or to identify particular organizational perspectives which may influence the program. Existing relationships among team members are an asset for the program management and should be encouraged whenever possible, especially at lower levels within the organization where the "real work" gets done. Clear indications of responsibility for program reviews and for "signing off" procedures to move to the next program milestone should be identified early. Furthermore, mechanisms for conflict resolution (i.e., disagreements among parent agencies or within the staff that adversely affect the program) should be available to program management, the earlier the better to minimize potential stress on program execution.

Requirements Management

This element may be the most difficult and time-consuming part of an IPO and should be managed effectively to minimize the natural tendency to have "requirements creep" in the program. "Requirements creep" refers to the tendency to add on additional requirements or "nice-to-haves" to a program as the program is underway. This can result in a perception of "gold plating" which will invariably invite increased administration attention and congressional scrutiny. Effective requirements management starts by understanding how each member organization of the IPO identifies user requirements, what metrics each organization employs to measure requirements success, and what process the provider organization uses to reach out to its customers. In DoD the process is well established, and provides rigorous traceability from requirements to capabilities. Similar processes do not exist in civil agencies, or if they do, they are much less rigorous or are driven by scientific processes that emphasize data collection and exploration rather than specific needs. Given the disparity among requirements processes, it is very important to identify and designate a preferred requirements process for the program in the starting MOU/MOA. Furthermore, this preferred process should include a requirements adjudication mechanism to minimize or discourage requirements creep by the partner agencies. Recognition also needs to be made of the greater than expected amount of time necessary for the program staff to deal with this issue that could have a direct bearing on program schedule and cost.

Funding Stability

Funding stability will be critical to the success of the IPO, therefore, mechanisms must be established in the early program planning stage to determine participating agency goals and interests, funding processes and schedules, and cost sharing arrangements before the program becomes a formal reality. The nature and type of a funding strategy that contributes to program stability will depend on whether the IPO is a technology demonstration program that will be quickly transitioned to a fully fielded capability, or a more traditional acquisition program for developing and procuring a large number of operational systems. This funding strategy should encompass the lifetime of the program and be agreed to by the participating organizations. Particularly important are penalties imposed on partner organizations for withdrawing from the program at unexpected times that would adversely affect program success, and at minimum, consideration of contingency funding sources and plans should that withdrawal occur regardless. As noted earlier, this action taken by a partner organization with far more resources than another partner can doom a program because of the inability of the other partner to compensate. The management team needs to take all considerations into account and make the program executive council aware of potential problems or concerns in time to influence or deter them.

Customer Responsiveness

Ensuring stakeholder support is another critical element to IPO success, and is complicated by the multiplicity of stakeholders influencing an IPO, some with conflicting goals and objectives. A leading example of this is the complication of additional congressional oversight committees and staffs that occur when considering IPOs that cut across civil and national security sectors. One approach to raising the level of customer awareness and support and to ensuring program awareness of differing agency perspectives is to involve them in the program from the earliest planning phases. This is a key part of the ACTD process, an example of the Executing Agent approach. ACTDs represent opportunities to demonstrate advanced technologies to military forces in the field, thus encouraging experimentation prior to full scale acquisition and development. By involving operational commands in the ACTD process, and making military utility a key benchmark for ACTD success, this increases customer familiarity with and support for the program. The IPO should have military or overall national security utility as an essential element of measuring program success in order to ensure stakeholder support. But the necessity of dealing with a wider range of customers by an IPO will complicate the ability of the IPO leadership to address this easily.

Another aspect to customer responsiveness is the necessity to keep external interested parties such as other parts of the administration and Congress informed and aware not only of program successes, but also of potential or impending problems and concerns. While no one likes to get bad news, keeping the senior leadership routinely informed can help to minimize potential surprises later. The executive council can be used to help the program management in this regard. The requirement for routine communication with all IPO participants by the program manager and deputy program manager may influence the kinds of skills and experience needed for these positions, more so than in a traditional acquisition program. For example, individuals with direct experience and knowledge of acquisition and program processes in multiple agencies, especially those in different sectors (i.e., military, intelligence, civil), should be an asset.

Cultural Alignment

Organizational culture can influence both operational capabilities and organizational structure and process. By its nature an IPO will have multiple cultures represented in its organization and personnel. Some cultures may be similar insofar as they have similar institutional objectives and experiences. Others may be radically different, such as IPOs formed from military or intelligence organizations and civil or science-oriented organizations. Organizational culture may also influence program control in terms of information flows and command hierarchies. Recognizing the effect of these different cultures on IPO organizational structure, management, and execution is crucial in order to understand and deal with internal bureaucratic behavior and with external pressures coming from not only the parent organizations but also other interested parties. Whether it is necessary to develop an overarching IPO culture may depend on the specific situation and duration of the program. In the Independent Agent approach, an overarching culture is required as a way to enable the staff to see how they as individuals and their position within the organization contribute to overall organizational success. Clear statements of IPO goals and expectations are important to ensure the staff understands the criteria for mission and program success.

Staffing

Last, but not least, is staffing. Once the IPO leadership and parent agencies have identified goals and management objectives for the program, they need to carefully think through how they will obtain the most qualified and experienced people to support the IPO. If the IPO is military in nature, the IPO leadership will have to account for a fair amount of staff turnover at routine intervals. Turnover has a direct bearing on institutional memory, but an IPO can benefit by having staff from one of its partner agencies come from a community where staff longevity is routine. The NPOESS program is an example of this situation where institutional memory resides more in the civilian employees and contractors than in the DoD side. Requirements for staffing success will include whether the IPO management identifies incentives to encourage new staff and retain existing staff, and if contingency plans are developed to account for the loss or transition of key staff members to other positions outside the IPO. Ideally, staff should view the IPO as an exciting place to *be*, but also as an exciting place to be *from*. Additional factors to be considered include IPO-required training and education, and post-IPO promotion opportunities.

Acknowledgments

This study could not have been undertaken without extensive support from many individuals and organizations throughout the U.S. national security space community. The authors are particularly grateful for the guidance and insights provided by Colonel Gary R. Harmon, USAF, Acting Director of the Air Force-NRO Integration Planning Group (ANIPG), Lieutenant Colonel Richard C. Einstman, USAF, and Jorn Kluetmeier, Aerospace Corporation, of the ANIPG, and also for insights and support provided by other members of the ANIPG. The authors are especially appreciative of the time and support to this study provided by the leadership and staff of the NPOESS Integrated Program Office, especially Colonel John Cunningham, USAF (ret.), Systems Program Director, and Bruce Needham, Director of Operations.

The authors are indebted to their RAND colleagues Jeff Drezner and Steve Berner for their critical reviews of the final document, and to Jeff Isaacson, Gene Gitton, Kevin O'Connell, Scott Pace, Michael Thirtle, Kenneth Reynolds, and Dave Frelinger for overall support and guidance.

To all of these people we are indebted, and any errors, oversights, and statements of opinion are those of the authors alone.

Glossary

ACAT	Acquisition Category
ACC	Air Combat Command
ACTD	Advanced Concept Technology Demonstration
ADA	Associate Director for Acquisition (NPOESS)
ADO	Associate Director for Operations (NPOESS)
ADTT	Associate Director for Technology Transition (NPOESS)
AFMC	Air Force Materiel Command
AFOTEC	Air Force Operational Test and Evaluation Center
AFSCN	Air Force Satellite Control Network
AFSPACE	Air Force Space (USSPACECOM component)
AFSPC	Air Force Space Command
AIA	Air Intelligence Agency
AIP	Aerospace Integration Plan (USAF)
AITF	Aerospace Integration Task Force (USAF)
ANIPG	Air Force-NRO Integration Planning Group
AOA	Analysis of Alternatives
ASD(C3I)	Assistant Secretary of Defense (Command, Control, Communications, and Intelligence)
ASJPO	Arsenal Ship Joint Program Office
ASN(RD&A)	Assistant Secretary of the Navy (Research, Development, and Acquisition)
AVHRR	Advanced Very High Resolution Radiometer
CAIV	Cost as an Independent Variable
CINC	Commander-in-Chief
CIO	Central Imagery Organization
CJCS	Chairman, Joint Chiefs of Staff
CMP	Convergence Master Plan (NPOESS)

CONOP	Concept of Operations
COTS	Commercial off-the-shelf
CRADA	Cooperative Research and Development Agreement
CSAF	Chief of Staff of the Air Force
DAB	Defense Acquisition Board
DARO	Defense Airborne Reconnaissance Office
DARPA	Defense Advanced Research Projects Agency
DASN	Deputy Assistant Secretary of the Navy
DCI	Director Central Intelligence
DDPO	Defense Dissemination Program Office
DESA	Defense Evaluation Support Activity
DIA	Defense Intelligence Agency
DM	Demonstration Manager
DMA	Defense Mapping Agency
DMSP	Defense Meteorological Satellite Program
DoC	Department of Commerce
DoD	Department of Defense
DoDD	Department of Defense Directive
DOE	Department of Energy
DSMC	Defense Systems Management College
DT/OT	Developmental test and operational test
DUSD(AT)	Deputy Under Secretary of Defense (Acquisition and Technology)
EMD	Engineering and Manufacturing Development
EOS	Earth Observing System
EPA	Environmental Protection Agency
EROS	Earth Resources Observation Systems
ESA	European Space Agency
ESOC	Environmental Satellite Operations Center
EXCOM	Executive Committee (NPOESS)
FEMA	Federal Emergency Management Agency

FGDC	Federal Geographic Data Committee
FIA	Future Imagery Architecture
FOIA	Freedom of Information Act
GA-ASI	General Atomics Aeronautical Systems, Inc.
GFE	Government Furnished Equipment
GIS	Geographical information systems
GOES	Geostationary Operational Environmental Satellite
GPS	Global Positioning System
IORD	Integrated Operational Requirements Document
IPL	Integrated Priority List
IPO	Integrated [or Interagency] Program Office
IPT	Integrated Product Team
ISSP	International Space Station Program
JARC	Joint Agency Requirements Council (NPOESS)
JARG	Joint Agency Requirements Group (NPOESS)
JCS	Joint Chiefs of Staff
JPO	Joint Program Office
JROC	Joint Requirements Oversight Council
JSF	Joint Strike Fighter
LRIP	Low-Rate Initial Production
MAE	Medium Altitude Endurance
MAIS	Major Automated Information System
MAISRC	Major Automated Information System Review Council
MC&G	Mapping, charting, and geodesy
MDA	Milestone Decision Authority
MDAP	Major Defense Acquisition Program
MITI	Ministry of International Trade and Industry (Japan)
MNS	Mission Needs Statement
MOA	Memorandum of Agreement
MOU	Memorandum of Understanding

NAF	Numbered Air Force
NASA	National Aeronautics and Space Administration
NAVAIR	Naval Air Systems Command
NAVSEA	Naval Sea Systems Command
NAVSPACECOM	Navy Space Command
NCA	National Command Authority
NDRI	National Defense Research Institute (RAND)
NEC	National Economic Council
NESDIS	National Environmental Satellite, Data, and Information Service
NIIRS	National Imagery Interpretability Rating Scale
NIMA	National Imagery and Mapping Agency
NLSRDA	National Satellite Land Remote Sensing Data Archive
NOAA	National Oceanic and Atmospheric Administration
NPIC	National Photographic Interpretation Center
NPOESS	National Polar-orbiting Environmental Satellite System
NRO	National Reconnaissance Office
NRP	National Reconnaissance Program
NSC	National Security Council
NSDI	National Spatial Data Infrastructure
NSpC	National Space Council
NSTC	National Science and Technology Council
OM	Operational Manager
OMB	Office of Management and Budget
OSD	Office of the Secretary of Defense; Office of Systems Development (NPOESS)
OSO	Operational Support Office (NRO); Office of Satellite Operations (NPOESS)
OSTP	Office of Science and Technology Policy
OTA	Other Transaction Authority

PEO	Program Executive Officer
PEO(CU)	Program Executive Officer for Cruise Missiles and Unmanned Aerial Vehicles
PM	Program Manager
POES	Polar-orbiting Operational Environmental Satellite
PPBS	Planning, Programming, Budgetary System
PSA	Principal Staff Assistant
R&D	Research and Development
SAE	Service Acquisition Executive
SAR	Synthetic Aperture Radar
SCD	Ship Capabilities Document
SecAF	Secretary of the Air Force
SMC	Space and Missile Systems Center (USAF)
SOCC	Suitland Satellite Operations Center
SPD	System Program Director (NPOESS)
SOPS	Space Operations Squadron (USAF)
SRTM	Shuttle Radar Topography Mission
SSEIC	Space Station Engineering Integration Contract
SSTI	Small Spacecraft Technology Initiative (NASA)
SUAG	Senior Users Advisory Group
SWOT	Strength, weakness, opportunity, threat
T&E	Testing and Evaluation
TENCAP	Tactical Exploitation of National Capabilities
TIPT	Transition Integrated Product Team
TM	Technical Manager
TPED	Tasking, production, exploitation, dissemination
TTO	Tactical Technology Office
UAV	Unmanned Aerial Vehicle
USACOM	United States Atlantic Command (later Joint Forces Command)
USAF	United States Air Force

USARSPACE	**United States Army Space Command**
USASMDC	**United States Army Space and Missile Defense Command**
USD(A&T)	**Under Secretary of Defense (Acquisition and Technology)**
USEUCOM	**United States European Command**
USGS	**United States Geological Service**
USP	**Unit Sail-away Price**
USSPACECOM	**United States Space Command**
VLS	**Vertical launch system**
XM	**Transition Manager**

1. Introduction

Background and Context

This study is being conducted for the Air Force-NRO Integration Planning Group (ANIPG). The ANIPG was established as a means by which to explore potential opportunities for cooperative efforts between the National Reconnaissance Office (NRO) and the Air Force in space programs. The context for the ANIPG lies in two areas. First, the Air Force has been conducting an effort to develop institutionally into an integrated operational aerospace force supporting national security objectives more efficiently and effectively. At the same time, both the Air Force and the NRO are faced with challenges and opportunities to expand their support to the warfighter in an era characterized as one of "high demand, low density," i.e., increasing demand for the products and services that the USAF and the NRO provide with fewer resources and lesser budgets. This document seeks to provide the foundation analysis for the management of joint or interagency cooperative programs through the concept of an interagency program office, or IPO.

Integrating Air Force and NRO Programs

The backdrop for consideration of interagency programs, particularly between the NRO and the Air Force, can be found in a number of recent events and trends affecting the national security space community. These trends include:

- The increasing number of military contingencies worldwide requiring intelligence and space system support for both short term tactical needs and longer term requirements (i.e., gathering information on specific targets over a long period of time)
- The overall change in focus and increase in long term intelligence requirements worldwide
- Insufficient agency resources to be able to conduct major independent (i.e., single-agency) intelligence and space program funding and acquisitions
- The absence of a single overwhelming target of focus such as the Soviet Union, but rather, an increasing diversity of intensive regional conflicts and threats to information systems such as telecommunications networks and interconnected infrastructures

- Despite some financial setbacks, continued growth in international space activities and the commercial space industry, and the increasing reliance of the national security community on commercial off-the-shelf (COTS) technologies
- Increased scrutiny of space and intelligence programs by the Congress and the Administration
- Independent institutional efforts in the Air Force and the NRO to better integrate air and space capabilities in support of the warfighter and the national decision-maker

All of these trends have contributed to the rationale for examining areas of common interest between the NRO and the Air Force, particularly where ongoing system and technology developments could possibly be better served by undertaking joint or interagency programs. This is not only true of the Air Force and the NRO, but also of many other agencies such as the Department of Defense (DoD), the National Oceanic and Atmospheric Administration (NOAA), the National Aeronautics and Space Administration (NASA), and the Defense Advanced Research Projects Agency (DARPA), as evidenced by satellite programs such as National Polar-orbiting Operational Environmental Satellite System (NPOESS) and Discoverer II. In other cases, decisions to combine similar functions and activities into a new organization, and designate that organization the Executive Agent for the government for that function, have been the result of integration, although perhaps driven by congressional pressures, budgetary reductions, or Administration mandates. The most recent example of this approach is the National Imagery and Mapping Agency (NIMA), established in 1996. It is evident that a number of different ways exist to approach the goal of integrating or collaborating on common programs or activities; however, how they are implemented can lead to a successful program, or lead to many complications and potential program failure.

Defining "Integration"

Since the notion of an IPO represents a collaborative, integrative approach among two or more organizations to gaining a capability or accomplishing common goals, the notion of "integration" needs to be understood. Recent unpublished RAND research on integration of space capabilities into mainstream military operations has defined "integration" in the context of the academic literature on organizational and economic behavior and analysis of "stovepipes" as distinct and autonomous functional areas within an organization. Much of the literature argues that stovepipes within organizations lead to inefficiencies or suboptimization–that is, each stovepipe within the organization optimizes its performance relative to its own goals, not to higher organizational goals. While stovepipes in and of themselves may serve an important purpose, those same stovepipes may have little visibility into how their performance affects the whole. Organizational integration becomes "the process by which activities are formed, coordinated, or blended into a functioning whole."[1] Thus, aerospace integration involves establishing the process(es) by which discrete functional activities are formed, coordinated, or blended into a functioning whole, i.e., the efficient and effective operational delivery of aerospace capability to the user (whether a warfighter or a national decision-maker). These discrete functional activities have both separate and distinct performance goals, operating budgets, and information systems, yet they share common activities such as budgeting, personnel, and security. They compete for resources at the corporate level, but may also have some degree of process integration through the development of concepts of operations (CONOPs). Part of laying out an IPO involves identifying which activities will be shared by the parent organizations, which will be conducted by one rather than the other, and what are expected or preferred outcomes from the assignment of responsibilities. We will discuss this in greater depth in Chapter 3.

Since integration of common activities and interests between two or more agencies is the focus of this study, examining the thinking within the Air Force and the NRO regarding aerospace integration is necessary to provide additional context for the IPO concept.

Aerospace Integration in the Air Force

During the CORONA Fall (1996) meeting on long-range planning and the future of the Air Force, the senior Air Force leadership enunciated a vision, later documented as *Global Engagement*,[2] that stated that the Air Force would evolve to an "air and space force" and eventually to a "space and air force." More recently, the Air Force has changed the terminology of "space and air" to "aerospace" to reflect a more integrated approach and to perhaps be less divisive among all the communities within the Air Force. Furthermore, it has revised and updated its corporate vision, releasing *Global Vigilance, Reach and Power*[3] in June 2000. The new vision emphasizes the contributions the Air Force makes to national security through balanced aerospace capabilities. Regardless of terminology, some uncertainty has existed about how to fully integrate both

---

[1]Bruce A. Friedman, MD, "Radiology Management," November/December 1997, pp. 30-36.

[2]*Global Engagement: A Vision for the 21st Century* (Washington, D.C.: HQ, United States Air Force, 1997).

[3]*America's Air Force Vision 2020, Global Vigilance, Reach, and Power* (June 2000), found at http://www.af.mil/vision.

space systems and operations into the mainstream operational Air Force and still provide the necessary space services required as the primary Service provider and about what "integration" actually means. By definition, choosing an integration strategy that realizes *Global Vigilance, Reach and Power* (and the earlier vision, *Global Engagement*) means change—change to the institution, organizations, culture, and people who comprise the Air Force. The Air Force also faces another dilemma in the process of integration: as the Air Force fulfills both its "steward of space" role and its role as an organize-train-and-equip, aircraft-oriented combat force, it faces the "enterprise question." What is the enterprise of the Air Force? Addressing the Air Force enterprise is beyond the scope of this study, but understanding the context and motivation behind aerospace integration are central to this effort.

Just prior to the release of the new vision statement, the Air Force published a white paper on aerospace integration, *The Aerospace Force: Defending America in the 21st Century*,[4] wherein "aerospace integration" is defined as:

> the set of actions harmonizing air and space competencies into a full spectrum aerospace force and advancing the warfighting capabilities of the joint force. These actions are parallel, sequential, and mutually coordinated. They occur simultaneously in the areas of organization, training, and equipment that lead to or reflect changes in warfighting concepts, doctrine, resourcing, and culture. Aerospace integration actions can also include actions that incorporate and exploit capabilities made available from non-military or non-Air Force communities.[5]

Aerospace Integration in the NRO

The origins of the National Reconnaissance Office occurred in decisions made by the Eisenhower Administration in the early days of the Cold War to combine major elements of the U.S. intelligence community into a single organization. This organization would be responsible for developing overhead technical systems that could collect reliable intelligence about Soviet strategic forces.[6] The current charter for the NRO was signed on August 11, 1965, by the then Director of Central Intelligence, William Raborn, and Deputy Secretary of Defense, Cyrus Vance. The charter stated that the NRO was to be a "separate agency of the DOD" responsible for "the management and operation" of the National Reconnaissance Program (NRP). The NRO is funded through the NRP, part of the National Foreign Intelligence Program.

---

[4]Department of the Air Force, *The Aerospace Force: Defending America in the 21st Century...a white paper on aerospace integration* (Washington, D.C.: Department of the Air Force, Spring 2000).

[5]Ibid., p. 3.

[6]R. Cargill Hall, NRO Historian, "The NRO at Forty: Ensuring Global Information Supremacy," unpublished article, c. Spring 2000. The NRO celebrated its 40th anniversary on August 31, 2000.

Operation Desert Storm in 1991 marked a key event for the military space and intelligence communities, including the NRO, in terms of intelligence requirements, customers, and technology applications in support of military operations. Coupled with the collapse of the Soviet Union, pressures developed to make the intelligence community more responsive to the needs of the warfighter. In 1992, the NRO established the Operational Support Office (OSO) to address tactical military issues.[7] Other outcomes attributable to the experiences from the war included greater collaboration among the military Service TENCAP (Tactical Exploitation of National Capabilities), and the consolidation of imagery and geospatial information responsibilities and capabilities within a single government organization, NIMA. Besides supporting the military and intelligence communities, the NRO now provides support to U.S. government agencies responding to natural and man-made disasters, the drug war, diplomatic and peacekeeping activities.[8]

In concert with the Air Force effort to integrate its air and space capabilities, the Director of the NRO and the Assistant Secretary of the Air Force for Space,[9] Keith Hall, signed a charter establishing the ANIPG in early 1998 as a joint office with staff from both the NRO and the Air Force. The ANIPG reports to both the D/NRO and the Assistant Vice Chief of Staff for the Air Force. As was mentioned earlier, the ANIPG has conducted a number of workshops that identify and explore common areas of technical and procedural cooperation between the NRO and the Air Force. Some of the workshops have led to prototype hardware being fielded, while others have identified areas of common concern among agencies on issues like tasking, processing, exploitation, and dissemination (TPED) of intelligence information. In May 1999 the ANIPG hosted a workshop on joint programs at which many of the participants shared common experiences and insights. That workshop was the foundation for subsequent studies such as this one.

---

[7]Ibid., p. 5.

[8]Ibid., p. 6.

[9]As noted on the NRO home page, "the Assistant Secretary of the Air Force for Space also serves as the Director of the NRO. The Director of the NRO is appointed by the Director of Central Intelligence (DCI) and the Secretary of Defense after being confirmed by the Senate as the Assistant Secretary of the Air Force for Space. The Director reports to the Secretary of Defense who, in concert with the DCI, has ultimate management and operational responsibility for the NRO. The DCI establishes collection requirements and priorities for satellite-gathered intelligence." See http://www.nro.odci.gov/index1.html.

Considerations for Addressing Aerospace Integration Between the NRO and the Air Force

The process the NRO and the Air Force must go through for both internal institutional integration and external cooperative integrative activities can be a complicated one. The NRO and the Air Force begin the process by defining their respective enterprises: "what" their specific purposes are and then "how" will they carry out those purposes. The NRO's "what" is found in its purpose statement:

> As the 21st century approaches, the NRO is guided by its vision of being Freedom's Sentinel in Space: One Team, Revolutionizing Global Reconnaissance.
>
> The mission of the National Reconnaissance Office is to enable U.S. global information superiority, during peace through war. The NRO is responsible for the unique and innovative technology, large-scale systems engineering, development and acquisition, and operation of space reconnaissance systems and related intelligence activities needed to support global information superiority.10

As found in its mission statement, the Air Force's "what" is "to defend the United States and protect its interests through aerospace power."[11] Its "how" is its most recent vision statement, *Global Vigilance, Reach and Power*.[12] Depending on priorities set by both the Air Force and external stakeholders, such as the Office of the Secretary of Defense (OSD) and the Congress, an evolution to an integrated aerospace force can occur in a variety of ways. The Air Force's approach to integration is found in a series of high level documents, such as *The Aerospace Force* and the *Aerospace Integration Plan* (AIP). Formulated and coordinated by the Aerospace Integration Task Force (AITF), the AIP's purpose is to examine ongoing aerospace integration "baseline" activities that have already been implemented in the USAF, to review the vision for the future aerospace force, and to introduce new high-leverage actions which have the potential to further significantly the integration of air and space capabilities.[13] The AIP serves as the inputs to a number of key Air Force program guidance documents, including the *Air Force Strategic Plan* and the *Air Force Annual Planning and Programming Guidance*.

Integrating operational air and space capabilities within the Air Force, within the NRO, and on an interagency level between the two organizations, as well as providing effective support to accomplish national security objectives, are

---

[10]National Reconnaissance Office, "Who We Are," found at http://www.nro.odci.gov/index1.html.

[11]*America's Air Force Vision 2020, Global Vigilance, Reach, and Power*, op. cit.

[12]Ibid.

[13]Draft AIP, Introduction, p. 6.

complex tasks requiring considerable time. Aerospace integration has implications for national security space doctrine, concepts of operations (CONOPs), education and training, organizational structures, and information/decision-making within those structures. It also has implications for the leveraging of new relationships with the private sector, the outsourcing of certain functions to the private and civil sectors, manpower allocations as well as career progression, resource allocation, and so on. Integration also has implications for understanding how changes or decisions in one area affect changes or decisions in another area. These implications need to be considered with respect to future joint or interagency program concepts.

Study Objective and Approach

The immediate objective for this study is to provide an assessment of the national security policy issues and implications associated with interagency program acquisition. Among the issues examined are included the organization and management of such a system by an interagency program office (IPO), the regulatory environment, potential security issues and concerns, defense planning and programming issues, and the influence of congressional and other external organizations and "players." This report examines the IPO concept from an air and space or aerospace integration perspective, and employs a "lessons learned" case study approach by analysis of several examples of joint or interagency program office formulation and implementation. Two cases of particular interest are included as appendices to this document and consist of the Arsenal Ship and the NPOESS.

It should be noted that a preferred interagency program acquisition approach may vary depending on the type of program or concept being pursued. For example, possible program categories might include:

- Research and development (R&D) program
- Technology demonstration
- Operational demonstration
- Acquisition of one or a small number of spacecraft
- Acquisition or "block buy" of a larger number of satellites

Specific details regarding acquisition complexity, program management and control, requirements management, funding and budgetary management, staffing, and other considerations will ultimately be situation- or program-specific. Nevertheless, a number of insights can be observed from both general principles and selected case studies that may be useful when it comes time to actually set up an interagency program. This report uses the mechanism of a

"SWOT chart" (Strengths, Weaknesses, Opportunities, Threats)[14] with a set of common elements to identify illustrative questions to be considered when determining an appropriate IPO concept. The SWOT chart is further employed to evaluate alternative approaches to IPOs. Finally, "lessons learned" are developed based on reviews of past or ongoing programs or technology demonstrations, and similarities and differences between single agency and joint programs are highlighted for insights into Air Force-NRO integration activities.

Organization of This Document

This report is structured as follows. This chapter has provided the introduction and background to this study, including the environmental context for aerospace integration and considerations for interagency system acquisition. Chapter 2 develops a framework for analysis by identifying opportunities for integrated activities and joint programs. It introduces the mechanism of the "SWOT chart" to pose illustrative questions that should be considered when identifying opportunities for interagency programs. Chapter 3 examines possible approaches to implementing an interagency program concept. Six alternatives are identified and analyzed based on alternative constructs, such as organizational structure (single participant, multiple participants), policies and regulations (traditional, streamlined), and mission (R&D/demonstration, small system buy, large system buy, data buy). Specific examples are used to illustrate the alternatives approaches. Chapter 4 offers findings and recommendations. The appendixes examine two case studies of interagency program acquisition and management in greater detail. Appendix A addresses the NPOESS program, the interagency weather satellite program established between the National Oceanic and Atmospheric Administration, the Department of Defense, and NASA. Appendix B addresses the Arsenal Ship program which was an interagency effort between the Navy and DARPA. Finally, a bibliography is included.

---

[14] "SWOT" is a business school approach to assessing programs. The authors are grateful to Liam Sarsfield and Scott Pace for suggesting this approach.

2. Framework for Analysis

Overview

This chapter will examine the rationale for undertaking interagency programs. Using the mechanism of a "SWOT chart" (defined subsequently) and its elements, we will define and describe considerations necessary to think about prior to initiating an integrated program, experiment, or concept. One aspect receiving particular attention is the importance of stakeholder support, including congressional oversight. While not a "fatal flaw" to interagency programs, the complication of multiple committees and oversight requirements should be taken into account when considering initial concepts and program organization. The discussion in this chapter will provide the foundation for a more detailed examination of alternative approaches to interagency programs in Chapter 3.

Identifying Opportunities for Integrated Activities and Joint Programs

The impetus for conducting integrated activities or establishing a joint[15] or interagency program can come from a variety of sources, including political pressures, operational needs, legal requirements, or economic imperatives. Often the Congress expresses interest in or mandates joint programs as a way to avoid duplication of effort among the military Services. Some of the reasons behind the establishment of joint programs include:

- Providing a joint combat capability
- Improving interoperability among the Services or components and reducing duplication
- Reducing development and production costs
- Meeting similar multi-service requirements
- Reducing logistics requirements through standardization[16]

---

[15]"Joint" connotes activities, operations, organizations, etc., in which elements of two or more Military Departments participate. See *DoD Dictionary of Military Terms*, http://www.dtic.mil/doctrine/jel/doddict/

[16]Eller, Lt Col Barry A., *Joint Program Management Handbook*, 2d ed., (Ft. Belvoir, Virginia: Defense Systems Management College Press, July 1996), p. 3.

Opportunities for conducting coordinated or integrated activities, such as developing CONOPS, determining required capabilities, or developing integrated acquisition programs can also occur at different points in time and the process is iterative, as shown in Figure 2.1 below. This assumes overarching policies and objectives provide guidance to the different organizations that may be pursuing similar capabilities or activities. The mechanism by which the activities of these organizations are systematically made known to each other to create the opportunity to conduct a joint or interagency program may be a formal one, or it may be *ad hoc*. An example of a formal process involving dual-use programs, especially in space programs or C4ISR, is the National Space Council in the first Bush Administration that developed a process for decision-making

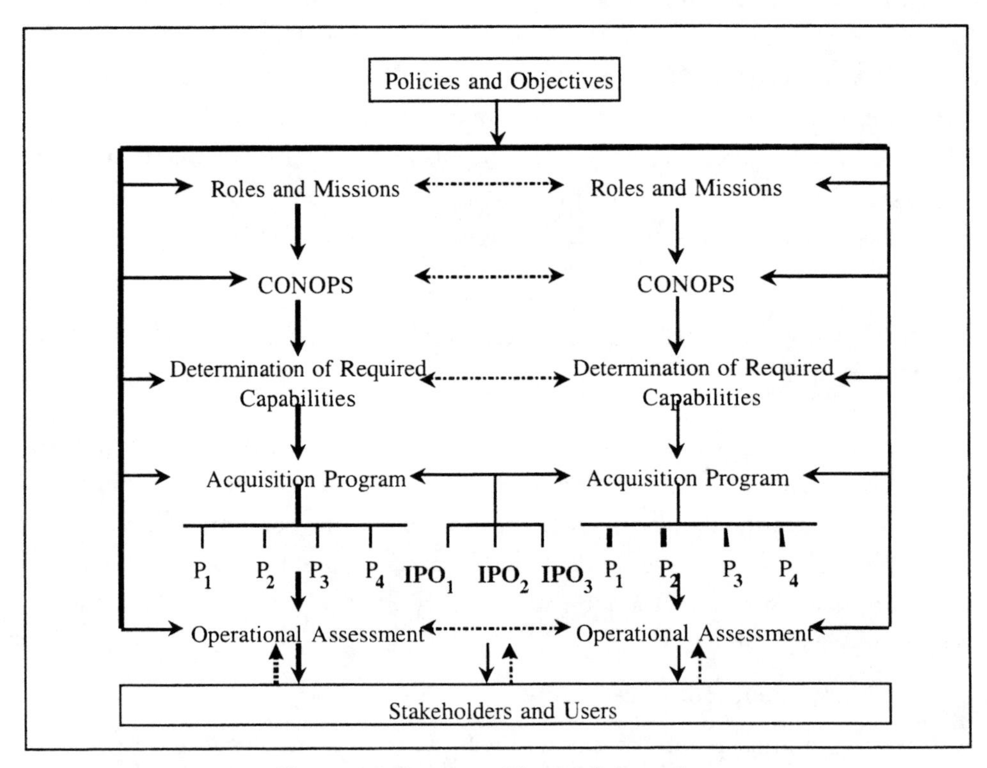

Figure 2.1 Opportunities for Integration

and adjudication of space-related issues and programs across agencies. The Clinton administration chose not to use this mechanism of space-related interagency cooperation and instead used other means.[17] Currently such a

[17]The National Space Council (NSpC) was first established in the Johnson Administration, and reactivated in the Bush Administration in 1989. In the latter administration the Council was chaired by the Vice President and supported by staff who worked closely with the agencies represented on the Council on cross-cutting or dual-use programs and issues. However, for various reasons the Clinton Administration chose not to continue the NSpC but instead chose to divide the policy implementation process among several White House organizations. These organizations include: the Office of Science and Technology Policy (OSTP), which oversees the National Science and

policy process mechanism like the National Space Council does not exist except through *ad hoc* or internal organizational measures.[18]

From a legal and regulatory sense, the *DoD Reorganization Act of 1986* (more commonly known as the Goldwater-Nichols Act) provides the legal foundation for joint programs and expands the role of combatant commanders in relation to the Service departments. Within the DoD process joint programs are governed by DoD Directive 5000.1, *Defense Acquisition,* and DoD 5000-R, *Mandatory Procedures for Major Defense Acquisition Programs (MDAPs) and Major Automated Information System (MAIS) Acquisition Programs,* which implements DODD 5000.1. These documents lay out the mandatory policies and procedures for the management of acquisition programs except when statutory requirements override.[19] Both documents were being revised as this report was being published. The NRO has its own acquisition series that is similar to the DoD Directive and regulations.

Important Considerations for Evaluating Alternative IPO Approaches

What are the key decisions in pursuing and implementing an integrated program concept and how should the NRO and the Air Force make those decisions? Furthermore, what criteria can be identified that measure progress towards successful integration, whether across institutions, in military and intelligence operations, or within an individual program? In an analytical sense, high-level functional decisions about what activities or programs to pursue via an IPO will drive organizational decisions about the development and implementation of IPO organizational structures. These decisions lead to cascading decisions about the management process guiding the IPO, the staff management process, and approaches to satisfying and maintaining external stakeholder support for the IPO program.

Technology Council (NSTC); the National Security Council (NSC); and the National Economic Council (NEC). OSTP advises the President and others in the White House on science and technology policies, plans, and programs of the Federal Government. The NSTC is the principal means for coordinating science, space, and technology research and programs within the Federal Government. The NSC advises and assists the President on national security and foreign policies, including military space policy, and coordinates these policies among various government agencies. While the NEC plays less of a role in space policy, its responsibilities for coordinating the economic policy-making process for domestic and international economic issues would likely touch on the health and status of the U.S. aerospace industrial base and international markets for U.S. commercial space activities.

[18]The ANIPG, for example, has hosted workshops on specific topics to identify similar programs or projects underway in different agencies and to determine areas of commonality or potential collaboration. Attendees have generally been from the defense, military Services, and intelligence communities and their contractors.

[19]Eller, op. cit., p. 11.

Using a "SWOT" chart approach with a set of common elements as an evaluation tool for the alternative IPOs, we have identified generic questions that should be kept in mind when making the initial decision to initiate a common program or experiment, and laying out the structure of an IPO. These elements include:

- Acquisition complexity
- Program management
- Program control
- Requirements management
- Funding stability
- Customer responsiveness
- Cultural alignment
- Staffing

Intended to be illustrative, the questions in each element may contribute to forming the basis for a programmatic "checklist," but answers to these questions can also contribute to establishing program metrics and ensuring that sufficient thought has been given in the early stages of a program to critical programmatic, organizational, and acquisition issues. Using these elements and questions will then enable us to compare the approaches for IPOs collectively in the next chapter and determine what works for an IPO and what does not, based on selected case study examples.

Before getting into the discussion of considerations, the elements in the SWOT chart need to be defined. "Acquisition complexity" denotes the degree of difficulty involved in acquiring a particular program or capability. It includes efforts to ensure that the program satisfies existing policy and objectives guidance and "front end" incentives to form the IPO. "Program management" refers to the organization, structure, and approach taken within a program to accomplish objectives. "Program control" is the ability to monitor and influence the operation of a program by the responsible individual, i.e., a program manager; this is also called "span of control." "Requirements management" involves the adjudication, coordination, and implementation of a common requirements process for the program. "Funding stability" is the process of maintaining funding support among the organizational partners over the lifetime of the program. "Customer responsiveness" refers to the program's relationship to its user base, e.g., how supportive are the users and stakeholders for the program, etc. "Cultural alignment" refers to the interaction of and implications for the program of the diverse organizational cultures inherited from parent or partner organizations. Finally, "staffing" is concerned with the staffing process of the program and the ability to attract qualified personnel to work in the program.

Acquisition Complexity

Understanding each contributing organization's vision, goals, and policies and where they both intersect and differ is crucial to providing the front end rationale for an IPO. In many cases the development of the memorandum of understanding (MOU) or memorandum of agreement (MOA) provides the foundation for the relationship between the parent agencies and the guiding document for the establishment of an IPO. Consequently, determining how this particular document is structured, what functional responsibilities are assigned within, and how the IPO is to be supported are critical to the future success of the IPO.

Illustrative questions to be raised include:

- How important is the proposed program to the missions of the parent organizations?
- What policies provide rationale for the IPO?
- What mission requirements does the IPO satisfy?
- Have incentives to form the IPO been identified?
- Have common agency goals been agreed upon?
- Are there any policies or regulations which guide/bind one agency but not the other?
- Are there any treaties or laws which affect the IPO or the program?
- How does the program support existing/future doctrine?
- Is there interoperability[20] with other existing or ongoing programs?
- How consistent/compliant with existing multi-agency planning guidance is the IPO?
- Have agency security concerns or issues been resolved, e.g., which agency will provide security support?
- Have mechanisms for information sharing among parent agencies and common responses to maintaining IPO information infrastructure assurance been agreed to and implemented?[21]

---

[20]"Interoperability" is defined in Joint Publication 1-02, *DOD Dictionary of Military and Associated Terms* as "(DOD, NATO) 1. The ability of systems, units or forces to provide services to and accept services from other systems, units, or forces and to use the services so exchanged to enable them to operate effectively together. "

[21]In *Sharing the Knowledge: Government-Private Sector Partnerships to Enhance Information Security,* Lieutenant Colonel Steven M. Rinaldi points out the differences among national security/military and intelligence communities with respect to the focuses of each sector, their respective information needs, and the speed at which they desire shared information. These differences contribute to a lack of common reporting criteria that satisfy the needs of all communities at all levels, in each organizational hierarchy, and at all times, and ultimately have an effect on the ability of the

Program Management

Once common visions, goals, and policies have been identified, these can provide the context and rationale for the organizational structure of an IPO. Desired end-state characteristics reflect the results of organizational integration within an interagency program. Organizational stovepipes are replaced by organizations integrated around operational capabilities. These integrated organizations have both the responsibility and authority to organize, train and equip operational capabilities such as remote sensing or C4ISR, no matter the source of origin: space-based, air-based, or hybrid system.[22] Applying appropriate management processes to the IPO is important and should be tailored to the specific program or project being considered for an IPO, as well as with the full understanding of the parent agencies' management processes. These management processes will include considerations of developing an acquisition strategy, a funding strategy, and a staffing strategy for the IPO and could be combined into an overall implementation plan.

Again, illustrative questions for this element could include:

- Has an MOU or MOA been developed/signed/implemented?
- Is the MOU/MOA sufficiently robust to ensure parent agency support?
- Have specific agency concerns about the purpose to the IPO been addressed satisfactorily in the MOU/MOA?
- Have cost sharing arrangements been satisfactorily addressed in the MOU/MOA?
- Are the agencies seen as partners or competitors?
- With regard to the program concept, are agency vision statements similar?
- If agency goals are dissimilar, are there some goals that can be agreed upon?
- Has an IPO implementation strategy been developed? Do the participants support the strategy?
- What is the larger policy context for the IPO implementation strategy?
- What is an appropriate organizational structure for the IPO?
- Does the organizational structure facilitate the goals of the IPO and its parent agencies?
- What is an appropriate chain of command for the IPO?

---

government to respond to information infrastructure threats. See Steven M. Rinaldi, *Sharing the Knowledge: Government-Private Sector Partnerships to Enhance Information Security*, INSS Occasional Paper 33 (U.S. Air Force Academy, Colorado: Institute for National Security Studies, May 2000), pp. 19-20.

[22]Remote sensing and C4ISR capabilities can be provided by aircraft, spacecraft, or some combination of platforms, depending on factors such as cost, availability, mission requirements, etc.

Program Control

Considerations may be made of how to encourage the uninterrupted flow of information and communication through an integrated organizational chain of command, with decisions being made with integrated information rather than piecemeal or relying on the decision-maker to integrate it. Approaches to the communications process within the IPO and with external entities need to be included in the management process. Maintaining information assurance in an era of the Freedom of Information Act (FOIA), privacy and confidentiality concerns, potential liability issues, and within national security guidelines is an area of growing importance to any program, but can potentially be greatly complicated by the nature of an interagency program.[23]

Questions to be raised for program control might include:

- Is IPO management inwardly focused (e.g., on intra-agency or internal IPO concerns) or outwardly focused (on the user)?
- Does IPO management draw on identified strengths of partner agencies sufficiently?
- Has an IPO acquisition strategy been developed?
    - Have the function and authority been assigned to a single agency? If so, how was the agency selected?
    - What is the acquisition process timeline?
- Does open communication exist throughout the IPO and its partner agencies?
- Has an executive council been established to support management of the IPO, provide expert reviews of the program, and/or provide "top cover" for the program?
- Have functional assignments been made for partner agencies?
- Are there existing institutional relationships among partner agencies that either enhance or conflict with the IPO?
- Are there existing contractual relationships with industry that will affect the IPO?
    - How flexible are those relationships?
    - What is the extent of certainty/uncertainty in the contracts?
    - What is the contractual governance structure?

---

[23]Ibid., pp. 26-41.

- What is the information process among organizations within the IPO? Does the organizational structure facilitate or hinder communication throughout the IPO?
- Are authority and responsibility collocated to encourage/allow for accountability within the interagency program?
- How are program reviews handled, and who participates in the reviews?
- What are "sign off" procedures for moving to the next program step?
- How are differences and disagreements adjudicated and resolved?

Requirements Management

Understanding each agency's approach to the requirements process, and identifying an approach to resolving procedural differences are also important at the onset, largely because of the potential time required later to adjudicate among differing or conflicting mission and system requirements. Also, developing mechanisms to deter or minimize "requirements creep" are necessary for effective requirements management.

Illustrative questions to be considered in this element are:

- Is there CONOPS compatibility, i.e., can the program satisfy multiple mission requirements?
- What is the process for adjudicating among the range of requirements offered by the parent organizations and their supporting communities (e.g., warfighters, scientists)?
- How are potential requirements conflicts resolved?
- Are requirements processes of the parent organizations understood sufficiently to identify common areas of interest and differing areas of potential concern?
- Is one requirements process designated at the onset as the program's preferred approach to requirements adjudication?
- How much time will be spent in coordinating and adjudicating requirements among parent agencies' user and stakeholder communities?

Funding Stability

Questions of ensuring support from parent agencies need to be addressed, especially a potential situation such as if one agency should withdraw its participation and support, or suffer through its own budget cuts that have an effect on the IPO's budget. For example, since the NPOESS budget comes from the DoD and NOAA, to the DoD its contribution to NPOESS is extremely small in comparison to other major programs like the Joint Strike Fighter (JSF) or a new carrier, while to NOAA the NPOESS is the largest program in its budget. Should the DoD cut its support to NPOESS, it would likely be very difficult if not impossible for NOAA to increase its budgetary contribution to the program to compensate. In addition, decisions about whether each agency pays in proportion to how its requirements contribute to overall costs versus sharing program costs equally over the life of the program need to be explicitly discussed early on. Measures to avoid continual expansion of requirements, especially when trying to create a broad network of agency and customer support, are also important to avoid escalating budgetary costs.

- Has an IPO funding strategy been developed?
- Are funding goals shared among partner agencies?
- Has sufficient thought been given to "What happens if..." one agency's funding goes away?
- Have contingency funding sources been identified?
- Have all management players been adequately apprised of the funding strategy? Are they supportive?

Customer Responsiveness

Ensuring stakeholder support will likely consume much of the IPO management's time and effort, probably at the expense of actual program management. While we do not believe this is a major flaw in the IPO approach, it is a big consideration for both program management and ensuring customer support. It may also have a greater effect on the kinds of skills required of the program manager and the deputy program manager in an IPO than in a single-agency program office. Furthermore, customer organizations' existing chains of command, authority, and responsibility need to be understood, as well as identifying other organizations that will have a stake in the outcome of the IPO program. This is largely because of the number of "players" in space and intelligence activities today, from the national security sector (including military and intelligence activities) to the civil sector. Stakeholders will include the parent agencies, the managing agencies, intelligence users (CINCs, imagery and geospatial analysts, etc.), operational users (e.g., USSPACECOM, NRO), and Congress. Within each of these groups, there are likely to be subdivisions among major commands (such as Air Combat Command [ACC], Air Force Materiel Command [AFMC], and Air Force Space Command [AFSPC] within the Air Force) and within commands (e.g., the space and missile communities within the broader USAF space community), with each having potentially conflicting as well as complementary interests. This part of the IPO management's responsibilities needs to be considered from the onset as a key factor in making the decision to pursue the IPO.

Figure 2.2 below understates the complexity of the situation, since it does not show two other sectors–the commercial and international sectors–that will also influence the behavior and relationships of the organizations shown therein. The figure illustrates several (Executive Branch) chains of command: operational, organize/train/equip, policy interests, and acquisition. Starting at the top, the President has multiple responsibilities as the NCA and the one who sets the overall policy tone for the national security community. The operational chain (i.e., CINC/warfighting) is shown in right-to-left diagonal crosshatching and flows down from the President through the Chairman, Joint Chiefs of Staff (CJCS) and down to the CINCs and their components. Of particular interest is United States Space Command (USSPACECOM) and its components, Army Space Command (USARSPACE), Navy Space Command (NAVSPACECOM) and 14th Air Force. USSPACECOM has responsibility for coordinating all military space requirements coming from the other theater and functional CINCs. Army Space Command is a major subordinate command of the Army's Space and Missile Defense Command (USASMDC). 14th Air Force serves as both a Numbered Air Force (NAF) to AFSPC and the Air Force Space (AFSPACE) component to USSPACECOM.

The Air Force "organize, train, equip" chain is shown in vertical lines. It flows through the Secretary of the Air Force (SecAF) and Chief of Staff, Air Force (CSAF) chains. The SecAF also has responsibility for the acquisition of forces; the acquisition chain is shown in dark gray. Air Intelligence Agency (AIA) has both a warfighting responsibility with respect to information warfare and an "organize, train, equip" responsibility, so it is shown in multiple-colored crosshatching.

A number of other government agencies have policy interests in decisions made by the NRO and the Air Force with respect to space system acquisitions and operations. Those entities are shown in light gray. Neither the NRO nor the Air Force can afford to ignore the implications that these other agencies and their actions have from a policy sense on the decisions they may make with respect to

Figure 2.2 Stakeholder Organizations: Macro View

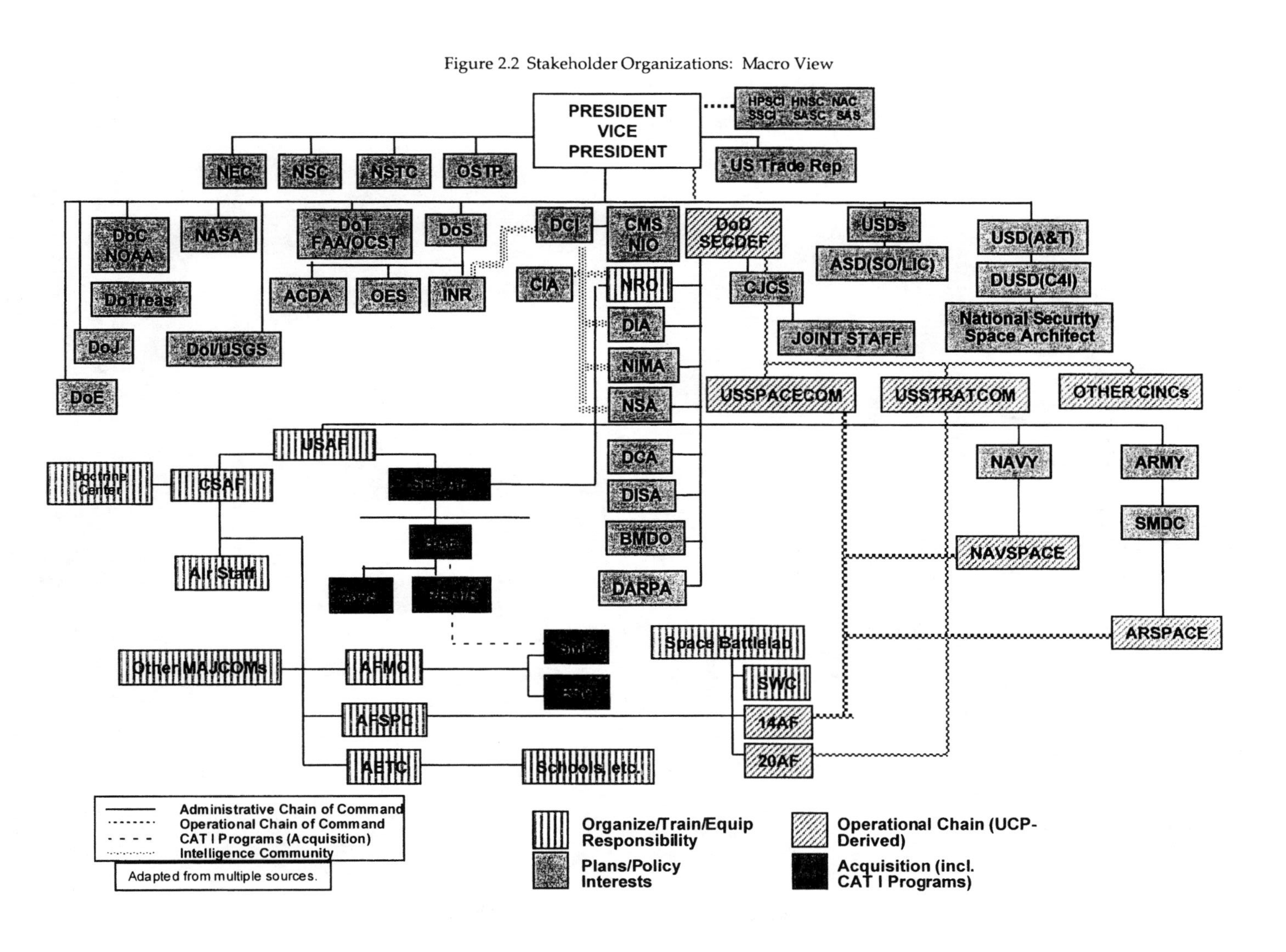

aerospace integration, such as budget and manpower reallocations and organizational change. Understanding the perspectives of these other agencies will also help maintain external stakeholder support for an IPO.[24]

Illustrative questions for this element may include the following:

- Who are stakeholders in the IPO? What relationship exists with the IPO?
    - Administration: budgetary oversight organizations?
    - Congressional: which committees?
    - Public sector: which government agencies (in addition to partners)?
    - Commercial: industry partners?
    - International: comparable programs?
- What is the extent of oversight?
- What is the oversight process?
- When are resources committed to the IPO establishment?
- Are key documents, e.g., MOA, sufficiently specific to encourage trust among stakeholders?
- Do means exist to continually address and satisfy agency concerns?
- How much time does IPO management devote to "educating" stakeholders and selling the program?
- Does the IPO have stakeholder "champions?"

Congressional Oversight

Because of the nature of program oversight provided by a number of congressional committees and subcommittees, we will address this area in greater depth.

---

[24]This report was largely completed prior to the release of the report of the Commission to Assess United States National Security Space Management and Organization (otherwise known as the "Space Commission") headed by Donald Rumsfeld prior to his nomination as Secretary of Defense in the Bush Administration. The Commission recommended a number of organizational changes, including that Title 10 U.S.C. should be amended to assign the Air Force responsibility to "organize, train and equip for prompt and sustained offensive and defensive air and space operations," (emphasis in original) and that the Secretary of Defense should designate the Air Force as Executive Agent for Space within the DoD. The Commission also recommended a realignment of Air Force and NRO programs, i.e., assign the Air Force Under Secretary as Director, NRO, and designate the Under Secretary as the Air Force Acquisition Executive for Space. See *Report of the Commission to Assess United States National Security Space Management and Organization*, January 11, 2001.

The U.S. Constitution, Article I, Section 8, gives Congress the authority to review government operations and administration. The congressional oversight function is exercised largely through the power to appropriate money and authorize its spending[25] and is conducted through hearings, investigations, and reports. The reasons for congressional oversight include:

- Determining whether the law is being executed and the money is being spent as appropriated
- Assessing whether conscientious efforts are being made to ensure limited resources are utilized in support of national security, domestic interests, and implementation of other policies
- Focusing on the Administration's failures (most likely) or accomplishments (less likely)
- Protecting and supporting favored policies and programs
- Asserting congressional authority in a particular area
- Encouraging Administration action through committee initiative or assertiveness (i.e., "spur things on")[26]

The timelines for congressional oversight generally occur after the President's Budget is submitted to Congress in February of each year. This phase in the resource allocation process is called "enactment" and covers congressional review of the Budget through hearings and the passing of legislation. It takes about nine months to transpire and ends when the President signs the annual authorization and appropriation bills. "Authorization" approves programs and specifies maximum funding levels and quantities of systems to be procured, whereas "appropriations" provides the budget authority with which to incur obligations and expend and outlay funds.[27] The third phase in the resource allocation process is termed "apportionment" and occurs following the President's signature on the authorization and appropriation bills and the release of funds by the Office of Management and Budget (OMB). The "execution" phase occurs when government agencies like the DoD obligate and expend funds for programs.[28]

---

[25]Wilbur D. Jones, Jr., *Congressional Involvement and Relations: A Guide for Department of Defense Acquisition Managers,* 4th ed. (Fort Belvoir, Virginia: Defense Systems Management College Press, April 1996), p. 101. Numerous overviews of the role of Congress in oversight of government programs have been written, but this document is one of the best sources for the defense acquisition manager found in our literature search.

[26]Ibid., pp. 102-103.

[27]C. B. Cochrane and G. J. Hagan, *Introduction to Defense Acquisition Management,* 4th ed., (Fort Belvoir, Virginia: Defense Systems Management College Press, June 1999), p. 69.

[28]Ibid., pp. 69-70.

All of these reasons and the approaches during each of the resource allocation phases that Congress takes to exercise its oversight responsibilities, mean that program managers must make every effort to understand the perspectives of the committees and subcommittees, to work with them, yet not be viewed as having too close a relationship, and to be as forthcoming as possible. This is difficult enough when a DoD or NRO program has to respond to only one set of committees, but it is compounded with an IPO that cuts across sectors, e.g., the armed services and intelligence committees. Later in Chapter 3 we will use a number of case studies to illustrate this point. Also, Appendix A, the NPOESS case study, for example, goes into much greater depth on this topic.

Tables 2.1 and 2.2 below provide tabular information on congressional committees in the House of Representatives and the Senate. It is not only important to know which Congressmen and Congresswomen serve on certain committees and subcommittees, but also how many and which congressional staff are affiliated with the committees since much of the interaction with committees will be at the staffer level. The tables also illustrate the jurisdictions of the committees. The NPOESS program, for example, has both houses' armed services committees as well as the Commerce and Science committees providing authorizing oversight, and the House and Senate Appropriations Committees providing appropriation for programs.[29]

Cultural Alignment

Culture can influence operational capabilities through the kinds of personnel, and their expertise and experience, who are assigned to the interagency program, as well as organizational structure and process. There are both practical benefits and consequences of cultural alignment, and analysis of organizational culture can help to clarify alternative institutional and management choices available to IPO leaders and to strengthen the cultural integration process. In conjunction with vision and policy statements, an integrated approach to organizational structures will encourage the development of an overarching IPO culture within which individuals as well as offices within the IPO will have a better understanding of their roles and contributions to IPO success. Developing an overarching organizational culture is directly linked to the organization's desired end state and objectives, and is important for establishing accepted performance and organizational metrics. Recognizing the inherent difficulty in attempting to

[29]The tables illustrate the breakout of committees in the fall of 2000 prior to the national election. As this document goes to publication, the congressional committee structure and membership were being revised.

Table 2.1 Congressional Oversight: House of Representatives

House of Representatives					
Committee	Number of Members		Number of Staffers		Description
	Rep. (Maj.)	Dem. (Min.)	Rep. (Maj.)	Dem. (Min.)	
Appropriations	31	27	16	17	Appropriate revenue for support of government
- Commerce/Justice/State, Judiciary, related	7	4	4	0	Jurisdiction over Dept. of Commerce, Competitiveness Council, Arms Control and Disarmament Agency, US Information Agency, and others
- Defense	10	6	22	0	Jurisdiction over DoD, OSD, Defense Agencies, CIA, Intelligence Community Staff
- Military Construction	8	5	3	0	DoD construction
Armed Services	32	28	51		Jurisdiction for Common Defense generally, Tactical intelligence & related activities, Scientific R&D, and Strategic and critical materials necessary for defense
- Military Procurement	15	13			Annual authorization for procurement of military weapons systems and related components
- Military R&D	15	13			Annual authorization for military R&D and related legislative oversight
Commerce	29	24	58	23	Interstate and foreign commerce generally
- Oversight and Investigation	9	7			Oversight of agencies, programs & departments within jurisdiction of full committee
Government Reform	24	19	31	29	Budget and Accounting, other than appropriations
- Government, Mgmt, Info., and Technology	6	5	7	0	Budget and accounting measures generally
- National Security, Vet. Affairs, and Int'l Relations	10	8	7	0	All matters relating to national security, VA, and international relations, and intelligence gathering activities
Science	25	22	15	9	All non-military scientific R&D
- Space & Aeronautics	15	13	6	1	Non-military space, especially NASA, Space Commercialization, and Earth remote-sensing
- Technology	11	9	5	3	Legislative authority on all matters relating to competitiveness, non-military R&D
Intelligence – Permanent Select	10	8	16	7	Intelligence community and the Director, CIA; authorizations for CIA, DIA, NSA, and all intelligence-related activities in DoD, State, and FBI
- Technical and Tactical Intelligence	6	4			

Table 2.2 Congressional Oversight: Senate

Senate					
Committee	Number of Members		Number of Staffers		Description
	Rep. (Maj.)	Dem. (Min.)	Rep. (Maj.)	Dem. (Min.)	
Appropriations	15	13	13	5	Appropriation of revenue for the support of the government
- Commerce/Justice/ State, Judiciary, related	6	5	5	2	Jurisdiction over Dept. of Commerce, Competitiveness Council, Arms Control and Disarmament Agency, US Information Agency, and others
- Defense	9	8	10	2	Jurisdiction over DoD, OSD, Defense Agencies, CIA, Intelligence Community Staff
- Military Construction	4	3	2	2	DoD construction
Armed Services	11	9	34	24	Air and space activities associated with DoD, as well as the common defense generally
- AirLand	5	4			Much of Air Force activities, including budgets for R&D
- Emerging Threats and Capabilities	5	4			Policies and programs to counter emerging threats; Information warfare, DARPA, DTRA, and non-traditional DOD R&D programs
- Strategic	5	4			Jurisdiction of BMD, SPACECOM, NSA, DIA, BMDO, NRO, and NIMA
Commerce, Science, and Transportation	11	9	30	21	Non-military space R&D, Commerce generally, Science and engineering R&D
- Communications	9	8	4	0	Jurisdiction of FCC, COMSAT, INTELSAT, Telecommunications generally, and spectrum allocation
- Science, Technology, and Space	5	4	3	0	Jurisdiction of NASA, NOAA, OSTP, Federal R&D funding, Internet, International science, technology, and space, commercialization of space
Government Affairs	9	7	25	15	Budget and accounting measures, other than appropriations
- Int'l Security, Prolifer., and Federal Services	6	5	8	3	Deals with effectiveness of present national security methods
- Investigation – Permanent Subcomm.	7	5	15	4	Efficiency and economy of operations of all branches of government
Judiciary	10	8	36	11	Committee of the Judicial branch
- Technology, Terrorism, and Government Information	4	3	3	2	Laws related to government information policy, oversight of technology, trade, and licensing
Intelligence – Select	9	8			Legislative oversight over U.S. intelligence activities and programs

supplant the organizational values and experiences from the parent agencies, the IPO leadership may want to build upon those values and experiences and develop a new overarching culture that defines what "world class" means for its staff and organization. Linkages among identity, mission, vision, and core value statements will influence organizational culture and affect individuals' understandings of their roles and contributions to the IPO's goals. Decisions to change distinct cultural characteristics, such as doctrine, concepts of organization, or staff career paths, should not be undertaken without understanding the implications for the potential cultural signals sent throughout the agency about what is valued and what is important.

Questions to be considered in this element may include:

- Is there a common culture among parent agencies that will provide the basis for an IPO-specific culture?
- How does the organizational structure facilitate the development of an IPO culture?
- Are there existing enhancements or barriers that can facilitate or hinder the development of an IPO-specific culture?

Staffing

The IPO management team needs to carefully think through how they will obtain the most qualified and experienced staff for the IPO. Given the goals and policies, organizational structure, and management processes have been identified and/or put in place, an important part of the IPO management responsibility is ensuring a high quality of personnel with the right kinds of experience and expertise to carry out the program successfully. Furthermore, the program manager must think ahead and ensure that staff (particularly military personnel who will leave the IPO in 18 months to two years) have good promotion and job opportunities upon transferring out of the IPO. This is an area with the potential for being neglected, with the result that the organization will be unable to attract quality personnel in the future. Questions to be considered include:

- Does IPO management adequately address staff career management issues?
- Has an IPO staffing strategy been developed?
- What IPO-related education is required of staff?
- What IPO-related training is required of staff?
- What incentives exist to attract staff to the IPO?
- What disincentives exist, and how can they be addressed?

- What incentives exist for agencies to send their best people to the IPO?
- What promotion opportunities exist for staff after their IPO assignments?

In Chapter 3 next, we will identify possible approaches to implementing an IPO concept. We will evaluate these alternatives through the use of illustrative examples of recent or ongoing programs. The questions raised in Chapter 2 above will be applied in evaluating these approaches in Chapter 3. Finally, broad insights and observations for Air Force-NRO collaborative programs in aerospace integration will be identified and discussed in Chapter 4.

3.Possible Approaches to Implementing an Interagency Program Office (IPO) Concept

Overview and Methodology

Chapter 2 provided broad considerations that will now be applied to an examination of alternative generic IPO approaches. Recognizing there is no one single approach to interagency acquisition, we attempted to identify a reasonable set of potential alternative approaches, as shown in Table 3.1. The examples include identifying an *Executing Agent* to conduct an interagency program of both interest to many users and of short term duration (e.g, ACTDs), a *System Integrator* (building on system elements provided by multiple agencies or nations in a common program), an *Independent Agent* (i.e., consolidating functions within one organization responsible for acquisition), a *Confederation* (brought together for limited but challenging objectives), and a *Joint Program Office.* Another alternative approach, *Commercial Prime,* was included for completeness but not analyzed in depth. Nevertheless, we believe it represents a viable alternative approach to interagency acquisition and needs further study.

Table 3.2 further defines each alternative along the lines of different constructs, such as organizational structure (i.e., whether there is a single participant or multiple participants), policy and regulatory approaches (whether traditional or streamlined), and missions, i.e., whether the program is R&D or a demonstration, a small system buy (e.g., very limited numbers of spacecraft), a multiple system buy (e.g., "block buy" of large orders of spacecraft), or a data buy (e.g., buying imagery from a variety of government and commercial sources). This section will further describe the alternatives and provide examples from prior or ongoing programs to illustrate specific points. The "real world" examples tend to fall in certain alternatives, largely because some of the alternatives as we have defined them would likely be very difficult to implement. Consequently, we devote more attention to those at the expense of some of the others. Nevertheless, all offer interesting analytical aspects worthy of examination. Following the analysis of each alternative, individual elements of the SWOT chart mechanism are used as frameworks to compare and contrast the alternative approaches collectively. Again, these elements include:

- Acquisition complexity
- Program management

- Program control
- Requirements management
- Funding stability
- Customer responsiveness
- Cultural alignment
- Staffing

Table 3.1 Definition of Interagency Program Office Alternative Approaches

Type	*Description*
Executing Agent	Agency designated lead for technology demonstration, development, acquisition, and/or operation of program for common or multi-user needs
System Integrator	Joint venture partners build system elements with lead organization operating as integrator
Independent Agent	Creation of new, independent, functionally focused entity to acquire, execute, operate program
Commercial Prime	Government partner using commercial development vehicle
Confederation	Multiple entities form acquisition "alliance" to accomplish limited, albeit challenging, objectives
Joint Program Office	Single, integrated program independent of, but responsive to, parent organizations

Table 3.2 Comparison of Alternative Acquisition Approaches and Illustrative Examples

Alternative Acquisition Approaches	Alternative Constructs and Examples							
	Organizational Structure		Policies and Regulations		Mission			
	Single Participant	Multiple Participants	Traditional	Streamlined	R&D/ Demonstration	Small System Buy	Large System Buy	Data Buy
Executing Agent	ACTDs	ACTDs		ACTDs	ACTDs			
System Integrator		ISSP	ISSP			ISSP		
Independent Agent	NIMA		NIMA	NIMA			NIMA	NIMA
Commercial Prime	Multiple Examples							
Confederation		NSLRS-D	NSLRS-D					NSLRS-D
Joint Program Office		GPS N-POESS Discoverer II Arsenal Ship	GPS N-POESS	Discoverer II Arsenal Ship	Discoverer II Arsenal Ship	N-POESS	GPS	

Legend: ACTD = Advanced Concept Technology Demonstration; ISSP = International Space Station Program; NIMA = National Imagery and Mapping Agency; NSLRSD = National Satellite Land Remote Sensing Data Archive; GPS = Global Positioning System; NPOESS = National Polar-orbiting Environmental Satellite System.

Alternative Approaches

Alternative Approach 1: Executing Agent

The definition of this alternative approach is an agency or organization designated as the lead for a technology demonstration, development, acquisition, and/or operation of program for common or multi-user (i.e., multiple agency, warfighter) needs.

An example of this alternative approach is the Advanced Concept Technology Demonstration (ACTD). ACTDs have their genesis in recommendations made by the 1986 Packard Commission, which called for early operational testing of prototype systems as a means to improve military capability, as a basis for realistic cost estimates prior to full scale development, and as a means for reducing "red tape."[30] ACTDs represent opportunities to demonstrate advanced technologies, whether emerging or mature, to military forces in the field, thus encouraging experimentation prior to full scale acquisition and development. ACTDs enable warfighters to identify potential changes to doctrine and are more like research and development efforts rather than major systems acquisitions. Because they are intended to last no longer than three years, the projects are scoped to demonstrate military utility and system integrity quickly.[31] As described by the DoD, the goal is to provide a prototype capability to the warfighter and to support him in the evaluation of that capability.[32] Military utility is the key attribute or metric of the ACTD and represents a significant departure from more traditional acquisition metrics.[33] The DoD identifies several key criteria by which ACTD candidates are evaluated:

- *Response to user needs*: meeting operational requirements and increasing user familiarity through technology demonstrations in realistic and extensive military exercises; providing a residual operational capability for the warfighter as an interim solution prior to procurement of a capability
- *Maturity of technologies*: exploiting mature or near mature technologies in order to reduce time and risk involved in technology development, thus permitting early user demonstration at greatly reduced cost and schedule
- *Potential effectiveness*: potential or projected effectiveness must be sufficient to warrant ACTD consideration or the capability must address a need for which there is no suitable solution[34]

ACTD objectives consist of: (1) conduct meaningful demonstrations of the capability; (2) demonstrate or enable CONOPs; and (3) prepare the technology to transition into formal acquisition without loss of momentum, assuming positive demonstration of military utility.[35] At the conclusion of the ACTD operational demonstration, three potential outcomes may result:

---

[30]Lt Gen Carlson, Director, J-8, "ACTDs: 'J-8 Persepctive,'" briefing, 11 January 2000.

[31]Michael R. Thirtle, Robert V. Johnson, and John L. Birkler, *The Predator ACTD: A Case Study for Transition Planning to the Formal Acquisition Process*, RAND MR-899-OSD (Santa Monica, California: 1997), pp. xiii-xiv.

[32]DoD, "Introduction to ACTDs," website information, August 2000.

[33]Thirtle, Johnson, and Birkler, p. 4.

[34]DoD, "Introduction to ACTDs," op. cit.

[35]DoD, "Introduction to ACTDs," op. cit.

- The user may recommend acquisition of the technology and fielding of the residual capability remaining after the ACTD demonstration to provide an interim and limited operational capability;
- If military utility is not demonstrated, the project is ended or returned to the technology base;
- The user's need is met by fielding the residual capability of the ACTD demonstration and there is no need to acquire additional capability.[36]

The management approach to ACTDs fits the lead agent option in that each ACTD is managed by a lead military Service or agency (Executing Agent) and driven by the potential user sponsor(s), usually a unified command. The Joint Requirements Oversight Council (JROC) makes a recommendation to the Deputy Under Secretary of Defense (Acquisition and Technology) (DUSD[AT]) regarding the lead Service and user sponsor as part of its review of ACTD candidates. An oversight group comprised of the user and development communities and chaired by the DUSD(AT) acts as a decisionmaking body that can respond quickly to significant program issues requiring guidance or approval and as a means of communicating program progress among the key participating organizations.[37] Management mechanisms or tools include an "Implementation Directive," a statement of roles and responsibilities for all parties agreed upon and signed by the principal participating organizations, including the sponsoring user (a unified command), the lead Service or agency, the executing acquisition organization (Service Acquisition Executive [SAE]) providing funding and materiel elements for the demonstration, and representatives from the Joint Staff/JWCA and DUSD(AS&C) representing DUSD(AT); and an ACTD "Management Plan" that lays out key objectives, approach, critical schedule events, participants, funding, and transition objectives.[38] Measures of ACTD evaluation (measures of effectiveness and measures of performance) are clearly identified and signature of the Plan constitutes endorsement by the significant participating organizations. A streamlined acquisition approach is key to the ACTD and includes the use of Integrated Product Teams (IPTs), and acquisition initiatives such as Cost as an Independent Variable (CAIV), use of commercial standards and products, and contractor logistic support concepts, are employed whenever feasible.[39]

---

[36]DoD, "Introduction to ACTDs," op. cit.

[37]DoD, "Introduction to ACTDs," op. cit.

[38]DoD, "ACTD Guidelines: Implementation Directives," October 1999.

[39]DoD, "ACTD Guidelines: Management Plans," October 1999.

At the program management level, the ACTD Demonstration Manager (DM)[40] and the ACTD Operational Manager (OM) provide day-to-day direction and are responsible for preparing periodic reports to the Oversight Group and other reviewing authorities. Specific individuals for these positions are identified by name in the Management Plan and agreed to at the onset of the ACTD.[41] Also included in the Management Plan are funding plans that cover the ACTD through completion of the demonstration but not necessarily beyond that, unlike a more typical acquisition program. The intent is to provide flexibility to the Management Plan in order to achieve the objectives in a timely manner, albeit with the approval of the Oversight Group.[42]

Transitioning the ACTD from a demonstration of mature or emerging technologies into the more formal acquisition process, assuming significant military utility, requires consideration of whether to move to the development phase ("Engineering and Manufacturing Development" [EMD]) to explore additional development, or, if further development is not necessary, to move to the "Low-Rate Initial Production" (LRIP) portion of EMD. Some of the transition decision process involves consideration of the class of the ACTD in question, i.e., whether it is Class I (typically information systems involving small quantitites), Class II (weapon or sensor systems similar to systems acquired through formal acquisition), or Class III ("systems of systems" potentially involving multiple Program Executive Officers and military Departments).[43] Transition planning is governed by DoD Directives 5000 and 5000.2R. Considerations for transitioning to LRIP include:

- *Contracting strategy*: motivating contractors to provide best value from a life cycle cost-effectiveness perspective and transitioning into LRIP without loss of momentum
- *Interoperability*: ensuring the ACTD can interface with other systems on the battlefield
- *Supportability*: ensuring that the fielded systems can be supported cost effectively
- *Test and evaluation*: early and continuous involvement by the test community to support transition planning

---

[40]For those ACTDs after 1998, the DM was renamed the Technical Manager responsible for all aspects of planning, coordination, and direction of development activities. The OM is responsible for planning, execution, and reporting of the Military Utility Assessment (MUA), while a new position, the Transition Manager (XM) leads the Transition Integrated Product Team (TIPT) to address transition planning and preparation. Michael J. O'Connor, "Advanced Concept Technology Demonstration: ACTD Process Overview," briefing, undated.

[41]DoD, "ACTD Guidelines: Management Plans," op. cit.

[42]DoD, "ACTD Guidelines: Management Plans," op. cit.

[43]DoD, "ACTD Guidelines: Transition," October 1999.

- *Affordability*: assessing life cycle affordability and application of a CAIV strategy to continuously identify means of reducing costs
- *Funding*: choosing the proper strategy for obtaining resources necessary for acquisition
- *Requirements*: evolving to a formal Operational Requirements Document (ORD) that effectively captures "lessons learned" in realistic exercises with the warfighter
- *Acquisition program documentation*: defining and planning for required documentation necessary prior to the acquisition decision at the end of the ACTD[44]

Based on earlier RAND research analyzing ACTDs in general and the Predator program specifically,[45] a number of ACTDs were analyzed with the goal of identifying potential insights into their contribution to interagency program alternatives. The Predator ACTD offers many useful insights for interagency acquisition. The Predator was intended to fill the Tier II (Medium Altitude Endurance [MAE]) Unmanned Aerial Vehicle (UAV) niche, i.e., the UAV had to be capable of flying 500 nautical miles, remain on station over the target for at least 24 hours, lift a 400-500 pound payload, and fly at altitudes of 15,000 to 25,000 feet. Furthermore, it had to provide a National Imagery Interpretability Rating Scale (NIIRS) rating of 6 or better at 15,000 feet, and demonstrate the integration of the Synthetic Aperture Radar (SAR) system with a 1-foot resolution at 15,000 feet.[46] Although the organization and conceptualization of Predator occurred prior to the genesis of the ACTD program, it became the first ACTD program in FY 1995. Predator was managed by a separate MAE UAV office within the UAV Joint Program Office. The UAV JPO is organizationally located in the Program Executive Officer for Cruise Missiles and Unmanned Aerial Vehicles (PEO[CU]). The JPO has coordinating and decision-making authority over all non-lethal UAVs being developed by the military Services, DARPA, and the OSD.[47] As the program developed, Predator participated in a number of operational exercises and was deployed twice to Bosnia for support to U.S. European Command (USEUCOM) in Operation Joint Endeavor. As the operational manager for Predator, U.S. Atlantic Command (USACOM) was responsible for gathering lessons learned during the operational deployments and incorporating them into the system and the Predator CONOPS. This

---

[44]DoD, "ACTD Guidelines: Transition," op. cit.

[45]See Thirtle, Johnson, and Birkler, op. cit.

[46]These requirements were laid out in a memorandum titled "Endurance Unmanned Aerial Vehicle (UAV) Program," by the Under Secretary of Defense, John Deutch, to the Assistant Secretary of the Navy for Research, Development, & Acquisition, July 12, 1993. See Thirtle, Johnson, and Birkler, pp. 10-11.

[47]Thirtle, Johnson, and Birkler, p. 6.

participation by the operational user community represented a unique feature of the ACTD process.

From the U.S. Government side, organizational participants in the Predator ACTD included:

- DUSD/AT
- The UAV JPO
- USACOM
- USEUCOM
- Air Combat Command (USAF)
- Defense Advanced Research Projects Agency (DARPA)
- Defense Airborne Reconnaissance Office (DARO)
- Defense Evaluation Support Activity (DESA)
- Air Force Operational Test and Evaluation Center (AFOTEC)

General Atomics Aeronautical Systems, Inc. (GA-ASI) was the prime contractor based on its prior experience with Predator's predecessor program, the GNAT-750. Subcontractors included Boeing, Magnavox, Verstron, Amerind, Westinghouse, and Loral.

Per ACTD policy, Predator had two managers: the OM, USACOM, representing the operational community and responsible for assessing the military utility of the program; and the DM, which focused on the engineering and technical aspects of the program. The developing agency was the Navy PEO(CU) and the DM was initially a Navy captain, and later a USMC lieutenant colonel. At USACOM, an Army colonel was the OM. As RAND notes, the decisions for selecting the OM and DM by the Under Secretary of defense (Acquisition and Technology) (USD[A&T]) were based on assessments of which organizations had the right mix of personnel and expertise, and which were likely to use Predator in the future.[48] A total of 10-12 people were assigned to the Predator ACTD full-time. Funding was provided by DARO outside the component Planning, Programming, Budgetary System (PPBS), i.e., directly from DoD and not through the military Services. Within OSD, an oversight panel was established for the Predator program. The panel was chaired by DUSD(A&T), and participants included ASN(RD&A), DARO, Joint Staff (J2), USACOM/Predator OM, Army/DAMO-FDZ, Army/CECOM, Navy (N85), USEUCOM, Assistant

[48] Thirtle, Johnson, and Birkler, pp. 27-28.

Secretary of Defense (C3I), PEO(CU)/UAV JPO, Predator DM, Air Force/ACC, and Marines.[49] Five integrated product teams (IPTs)—systems engineering and integration, payloads and data-links engineering, operations and demonstration support, business management, and contracting—were set up within the Predator ACTD office and mirrored in the prime contractor office. A number of MOAs were developed by the PEO(CU) to establish working level relationships with outside organizations such as DESA, AFOTEC, and Naval Air Systems Command (NAVAIR). The MOAs, IPTs, and small organization reflected the streamlined management approach that characterized the Predator program.[50]

Innovative program controls were critical to Predator. Because of the streamlined nature of the program, fewer documentation requirements were placed on the contractors, which meant a high level of trust between the program office and the contractors had to be present. Furthermore, novel approaches to user training were employed to speed up the development of an effective training program and incorporate lessons learned from the operational deployments. A number of control and communication techniques were employed by the program office, including daily communication between the DM, OM, and the GA-ASI program manager, weekly program reviews, government and contractor staff meetings, quarterly DM program reviews to the PEO(CU), and periodic reviews by the OSD Oversight Panel.[51] User requirements were developed and updated throughout the program in CONOPS working groups sponsored by USACOM. These working groups increased user familiarity with the Predator, thereby expanding user support for the program.

Problems with transitioning the Predator from an ACTD program to an operational capability occurred at the point of designating a lead military Service for the program. RAND interviews and analysis indicated that the December 1995 decision by the JROC to make the USAF ACC the lead service led to program implementation issues. ACC had not been an active participant in the ACTD process and believed that the USACOM operational evaluation and CONOPs omitted three critical ACC operational requirements. ACC then

---

[49]Thirtle, Johnson, and Birkler, pp. 28-29.

[50]Thirtle, Johnson, and Birkler, pp. 29-30.

[51]Thirtle, Johnson, and Birkler, pp. 31-32.

incorporated these requirements into the formal development of the ORD. Determination of military utility was then in question because (1) no formal definition of military utility existed other than USD(A&T)'s declaration that "the user-sponsor is responsible for assessing the worth of an ACTD" and (2) ACC disagreed with the approach USACOM had taken in assessing Predator's operational utility through exercises and the Bosnian deployments.[52]

Insights gained from the Predator analysis are shown in Table 3.3 below. Additional insights regarding the characteristics of the program office, measures of program control, attributes of key personnel such as the DM, selection of a key Service, and other issues, are discussed in greater detail in the 1997 RAND report.

Table 3.3 Relevant Insights from Analysis of the Predator ACTD

• Given the necessarily fast pace of the ACTD process, confident, effective, and innovative individuals are critical to the success of a program.
• The lead service must be selected early in the ACTD process to ensure that (1) proper test and logistics planning occurs; (2) operational requirements are fleshed out, and (3) to ensure program longevity and success, warfighters have complete buy-in to the system, participating in the ACTD from start to finish and being stakeholders in the product, not just observers.
• An ACTD needs to be managed significantly differently than are formal acquisition programs, because of the (1) fast-paced program schedule, (2) small numbers of program office personnel, and (3) limited guidance on how to perform the acquisition of the system.
• The lead-service organization should develop a draft ORD during the ACTD process. The process of writing and constantly updating the ORD will (1) resolve any misunderstanding of requirements among developers and warfighters, (2) help define quantitative system specifications, and (3) facilitate transition of the ACTD to the acquisition process.
• ACTDs planning discussions must involve operational users, lead-service personnel, and acquisition experts who can assess functional areas such as test, logistics, engineering, and affordability. Such planning is especially important if a strong probability exists that the ACTD will make the transition to the formal acquisition process upon its completion.

Adapted from: Thirtle, Johnson, and Birkler, p. 78.

---

[52]Thirtle, Johnson, and Birkler, pp. 34-36.

Alternative Approach 2: System Integrator

This alternative occurs when joint venture partners build elements of a system or program with a lead organization operating as the system integrator. It maximizes the direct application of expertise from other organizations under the management of a single organization acting as overall system integrator. However, depending on the number of program partners, it can be potentially difficult to manage unless carefully structured (through MOAs, etc.) at the beginning of the program.

NASA's International Space Station program (ISSP) is an example of the System Integrator approach to conducting a joint program. The ISSP is touted as being the largest and most complex international scientific project in history. It is scheduled for completion in 2004[53] and is estimated to have a life span of up to 20 years. When completed it will have a mass of about 1,040,000 pounds, and will measure 356 feet across and 290 feet long, with almost an acre of solar panels to provide electrical power to six state-of-the-art laboratories. It will be in an orbit of 250 statute miles altitude with an inclination of 51.6 degrees that will allow visitation by the launch vehicles of all ISSP partners. This orbit also allows 85 percent of the world to be observed and overflight of 95 percent of the world's population.[54]

In this case, NASA is the lead agency responsible for coordinating the program at an international level. Sixteen nations have signed agreements establishing the framework for cooperation among the Space Station partners for the design, development, operation, and utilization of the ISSP. Led by the U.S. Department of State in January 1998, the 1998 Intergovernmental Agreement on Space Station Cooperation was signed by representatives of Russia, Japan, Canada, and the participating countries of the European Space Agency (ESA), including Belgium, Denmark, France, Germany, Italy, the Netherlands, Norway, Spain, Sweden, Switzerland, and the United Kingdom. Three bilateral memoranda of understanding were also signed on February 24, 1998 by the NASA

---

[53]It is the view of the NASA Advisory Council in their assessment, *Report of the Cost Assessment and Validation Task Force on the International Space Station*, April 21, 1998, that completion of ISS assembly is likely to be delayed from one to three years beyond December 2003. The ISSP has also been facing increasing budgetary scrutiny as this report was being written.

[54]NASA International Space Station website, undated.

Administrator separately with his counterparts from the Russian Space Agency, the ESA Director, and the Canadian Space Agency President. These agreements supersede previous Space Station agreements signed in 1988, and reflect changes to the ISSP resulting from significant Russian participation in the program and design changes made since the original partnership in 1993.[55] Since the initial agreements were signed, Brazil has also signed up for participation in the ISSP.

The elements provided by these other nations include:

- A 55-foot robotic arm to be used for assembly and maintenance tasks on the Space Station provided by Canada
- A pressurized laboratory, named *Columbus*, to be launched on the Space Shuttle and logistics transport vehicles to be launched on the Ariane 5 launch vehicle, provided by ESA
- A laboratory with an attached exposed exterior platform for experiments as well as logistics transport vehicles, provided by Japan
- Two research modules; an early living quarters called the Service Module with its own life support and habitation systems; a science power platform of solar arrays supplying about 20 kilowatts of power; logistics transport vehicles; and Soyuz spacecraft for crew return and transfer; all provided by Russia
- Additional equipment for the station provided by Brazil and Italy.[56]

The ISSP has undergone several evolutionary changes to arrive at an integration solution that NASA could control. Initially, NASA structured the program as a lead center effort. Since the ISSP effort consumed a major portion of the NASA budget, this led to one of NASA's field centers being abnormally large in comparison to other installations. A strategic shift was made to distributed "work package" elements (distributed to various NASA field centers, each with their own prime contractor) and a new Space Station Engineering Integration Contract (SSEIC) to coordinate the work packages and provide overarching systems engineering integration. The SSEIC concept proved unwieldy and was never fully accepted within the NASA culture. At significant cost to the government, SSEIC was cancelled and the government office that oversaw the contract closed. The work package concept was substantially retained and the ISSP was consolidated under a system integration function assigned to a single NASA center.

---

[55]"Partners Sign ISS Agreements," November 28, 1998, posted on NASA International Space Station website.

[56]"The International Space Station," overview, posted on NASA International Space Station website.

Though the ISSP history has been turbulent, the system integration office experiment has been shown to be effective in coordinating a large and diverse program with significant international participation. The private sector has also demonstrated an ability to accept a large percentage of the management burden of dealing with a joint program and has helped the government retain a smaller system integration staff.

However, concerns remain about the outcome of several critical issues: the program size, complexity, and ambitious schedule; the schedule uncertainty associated with Russian participation in the ISSP; and other critical risk elements having an adverse impact on the ISSP cost and schedule, such as hardware qualification testing, on-orbit assembly complexity, crew return vehicle development, multi-element integrated testing, U.S. laboratory schedule, training readiness, software development and integration, and parts and spares shortages.[57] The international participation has its own limitations due to internal governmental issues, funding and schedule commitments and adjustments, and ISS management requirements including partner approval of possible modifications to the assembly sequence, ground operations, and on-orbit operations. This situation has been especially critical because of questions about Russian involvement and ability to maintain and implement its partnership commitment to the ISSP. The uncertainties of Russian participation have driven up the costs of the U.S. participation and have significantly affected U.S. schedules and final designs.[58]

Alternative Approach 3: Independent Agent

This approach is characterized as "independent agent" and is defined as the creation of a new, independent, functionally focused organization to acquire, execute, and operate a program and to consolidate functions transferred from many agencies into a single entity. The positive attributes of this approach are that it facilitates organizational and funding focus on a single program or functional area, such as, for example, consolidating the government's imagery and geospatial information efforts into a single provider/supplier organization. On the other hand, the agency needs to be given the appropriate authorities and responsibilities to carry out its mission(s). Furthermore, potentially conflicting internal cultures inherited from its legacy agencies may influence the agency's organizational cohesion and affect its ability to meet its customers' needs effectively.

---

[57]NASA Advisory Council, *Report of the Cost Assessment and Validation Task Force on the International Space Station*, April 21, 1998.

[58]NASA Advisory Council, op. cit.

A prominent example of this alternative approach is NIMA which was created on October 1, 1996, by the then Deputy Secretary of Defense, John White, and the Director of Central Intelligence, John Deutch, from a number of other intelligence organizations within the national and military intelligence communities. These organizations included the Central Imagery Organization (CIO), Defense Mapping Agency (DMA), Defense Airborne Reconnaissance Organization (DARO), the National Photographic Interpretation Center (NPIC), the Defense Dissemination Program Office (DDPO), and the imagery exploitation and dissemination elements of the CIA, NRO, and the Defense Intelligence Agency (DIA). NIMA was established,

> to address the expanding requirements in the areas of imagery, imagery intelligence, and geospatial information. It is a Department of Defense (DoD) combat support agency that has been assigned an important, additional statutory mission of supporting national-level policymakers and government agencies. NIMA is a member of the Intelligence Community and the single entity upon which the U.S. government now relies to coherently manage the previously separate disciplines of imagery and mapping. By providing customers with ready access to the world's best imagery and geospatial information, NIMA provides critical support for the national decisionmaking process and contributes to the high state of operational readiness of America's military forces.[59]

The expectation was that NIMA would evidence a natural convergence of the mapping and image-exploitation functions, as each evolved into becoming more digitally oriented from the labor-intensive technologies of photointerpretation and map generation, into a single, coherent agency focused on the construct of a geospatial information system (GIS).[60] This convergence has yet to occur, given the difficulties inherent in merging distinct cultures and complicated by the growth in commercially available imagery for a wide range of uses beyond the national security realm.

Since its establishment, a number of official studies and commissions reviewed NIMA and made some common recommendations, such as strengthening NIMA's role as the functional manager for imagery and geospatial information and identifying the need for agile, integrated tasking and other capabilities across satellite, airborne, and commercial sources of imagery.[61] NIMA has multiple roles:

---

[59]See http://164.214.2.59/general/faq.html. Quoted in *The Information Edge: Imagery Intelligence and Geospatial Information in an Evolving National Security Environment: Report of the Independent Commission on the National Imagery and Mapping Agency*, (hereafter called NIMA Commission Report), December 2000, Final Report, p. 9.

[60]NIMA Commission Report, p. 9.

[61]NIMA Commission Report, p. 5.

- *Intelligence producer*: providing intelligence information through imagery analysis and photo interpretation to a wide variety of users and decisionmakers, from military commanders in the field to the President and the Executive branch
- *GIS provider*: providing mapping, charting, and geodesy (MC&G) and information to a wide variety of users at all levels
- *Acquisition*: unlike DMA and NPIC, which relied on their parent organizations for acquisition of systems and capabilities, NIMA must conduct system engineering and acquisition activities for which it does not have the institutional cadre or organic assets. This is an area of development within NIMA and is particularly important in light of expanding reliance on commercial providers for imagery and GIS information.[62]

NIMA also relies on its satellite developer and provider, the NRO, for imagery. Thus, the NRO is a supplier to NIMA, despite its longer history and greater funding. An issue of growing importance to both organizations is the acquisition of the next generation of imagery satellites and their associated ground equipment, a program called the Future Imagery Architecture (FIA).[63] As the NIMA Commission noted,

> For the first time, the design of an NRO system was dictated more by requirements and less by technology, and was "capped" in terms of overall system cost. As a consequence of the requirements versus technology change, it will end up delivering imagery, much of which could be acquired from commercial imagery providers whose technology is not far below that of the NRO. As a consequence of the funding cap, there are currently five capabilities validated by the JCS, which FIA will not provide. From the Commission's perspective, the phasing of FIA, which delays integration of airborne and commercial imagery into the "system," is suboptimal.[64]

NIMA is facing challenges to its ability to continue to provide leadership in the market for geospatially referenced intelligence analysis, both as the largest customer and the largest supplier of the digital source for intelligence products. The commercial sector finds it increasingly difficult to deal with NIMA's lengthy acquisition processes and legacy systems, and NIMA itself needs to foster innovative approaches to commercial acquisition and to staying current with advances in the commercial sector. Decentralization of information sources and providers runs counter to traditional hierarchical approaches to government acquisition. While NIMA has adopted a Commercial Imagery Strategy, it has been slow to implement it because of a lack of understanding about the

---

[62]NIMA Commission Report, pp. 17-19.

[63]Specific details on FIA are classified and will not be discussed here.

[64]NIMA Commission Report, pp. 47-48.

relationships between commercial imagery and classified imagery information, problems with implementing its procedures for purchasing commercial imagery, a high turnover in experienced people, insufficient funding, and a perception that commercial imagery should be purchased only for the raw image, not for the value added information and analysis of that image.[65]

Alternative Approach 4: Commercial Prime

The use of commercial prime contractors to execute government programs is certainly not new. Allowing a commercial firm, or set of firms, to perform the joint integration function does mean, however, that the government has less control over the program. This alternative would free each service to focus on the development of performance requirements that meet the warfighter needs they are most familiar with. The contractor would then be responsible for soliciting these requirements and integrating them into a set of performance specifications to guide trade studies aimed at delivering the highest-performance solution at the least cost and risk.

Such an approach would place the greatest amount of responsibility on the contractor of any of the alternative approaches RAND considered. Recent moves to increasingly favor commercial solutions and innovation, while laudable, can increase risk to the government. The incentive structure put in place by the government must be carefully crafted to avoid perverse behavior on the part of private firms. This alternative should be considered a high-risk option and is perhaps a poor match for the development of unprecedented military systems.

Because of the wide range of examples in this category, it is only included here for completeness. Additional research on this topic can be found in a number of other documents included in the Bibliography.

Alternative Approach 5: Confederation

Alternative Approach 5 is defined as multiple agencies forming an acquisition "alliance" to accomplish limited, but perhaps challenging, objectives. Program management may be conducted through a joint or combined program office. This approach provides a coordinated agency focus on and support for a specific issue or program. We will explore this approach in greater depth by looking at the Department of the Interior's establishment of a permanent government archive containing satellite remote sensing data of the Earth's land surface. The archive is called the National Satellite Land Remote Sensing Data Archive, and it

---

[65]NIMA Commission Report, pp. 55-56.

is a comprehensive, permanent, and impartial record of the Earth's land surface derived from about 40 years of satellite remote sensing.[66] Much of that data comes from the imagery provided by the LANDSAT program, and is archived, managed, and distributed by the USGS EROS (Earth Resources Observation Systems) Data Center in Sioux Falls, South Dakota. Other archived data comes from the Advanced Very High Resolution Radiometer (AVHRR) carried aboard the NOAA Polar-Orbiting Environmental Satellite (POES); and more than 880,000 declassified intelligence satellite photographs. In 2001 the planned archived holdings will include:

- LANDSAT 7 NASA's MODIS instrument, part of the Mission to Planet Earth's Earth Observing System;
- ASTER, a cooperative effort between NASA and Japan's Ministry of International Trade and Industry (MITI);
- the Shuttle Radar Topography Mission (SRTM), a joint venture between NIMA and NASA;
- LightSAR, a NASA synthetic-aperture radar instrument; and
- NASA's Small Spacecraft Technology Initiative (SSTI)[67]

By 2005 it is expected that the archives will hold about 2,400,000 gigabytes of data, an enormous amount of information that will be made available to a worldwide community of scientific users. The Archive is an important national resource and represents a huge investment requiring careful management.

The Archive has its regulatory basis in the Land Remote Sensing Policy Act, the U.S. National Space Policy, and OMB Circulars A-76 and A-130. Congress designated the Secretary of the Interior to provide for "the long-term storage, maintenance, and upgrading of a basic, global, land remote sensing data set and providing timely access to it."[68] The United States Geological Survey (USGS) manages the program in cooperation with a wide range of other Federal, state, and local government agencies and in partnership with the private sector. The National Satellite Land Remote Sensing Data Archive (NSLRSDA) Advisory Committee provides oversight of the Archive.

The Archive is considered to be a government function, but has a relationship to encouraging the commercialization of land remote sensing as a long term U.S.

---

[66]U.S. Geological Survey, EROS Data Center, *National Satellite Land Remote Sensing Data Archive*, overview, 11 February 1998, available at http://edc.usgs.gov/program/nslrsda/overview.html.

[67]Ibid.

[68]Department of the Interior, National Satellite Land Remote Sensing Data Archive Advisory Committee, Memorandum to the Secretary of the Interior, re: *National Satellite Land Remote Sensing Data Archive Policy White Paper*, January 25, 1999, found at http://edc.usgs.gov/programs/nslrsda/advisory/whitepaper.html.

policy goal. To that end, the NSLRSDA Advisory Committee writes in its white paper:

> The dynamics of the remote sensing industry, including the value-added data enhancement sector, are evolving, and will continue to evolve, dramatically. The National Space Policy establishes the need for EDC [the USGS/EROS Data Center] to assist in this evolution. It requires government agencies to support the development of U.S. commercial Earth observation capabilities by pursuing technology development programs, including partnerships with industry; ...providing U.S. Government civil data to commercial firms on a nondiscriminatory basis to foster the growth of the "value-added" data enhancement industry; and making use, as appropriate, of relevant private sector capabilities, data, and information products.
>
> To meet these obligations and objectives, it is recommended that EDC enhance its outreach program to the private sector. Examples of how this might be accomplished include establishing a goods and services clearing house for the value-added community; providing information regarding services available from the value-added community; holding annual meetings with the value-added and user communities; and/or providing a central web site that lists sources for products and services. Particular attention and assistance ought to be paid to smaller and new companies in order to assist the development of a diverse, competitive marketplace. It is particularly noted that, with the advent of Landsat 7, data costs to the public are expected to drop dramatically due to returning Landsat operations to the government. EDC should ensure that these lower costs actually result in significant distribution by making the new costs as widely known as possible.[69]

The relationship of the government-provided Archive to the emerging commercial remote sensing industry is a subject that is addressed at far greater length elsewhere.

The USGS has also initiated a number of MOUs with NOAA, NASA, DoD, the Environmental Protection Agency (EPA), and the Federal Emergency Management Agency (FEMA),[70] and has expressed interest in forming cooperative partnerships with organizations from all levels of government and industry for geospatial data production and mapping science research. Furthermore, it states that "[T]hrough a variety of partnership mechanisms, the USGS seeks to ensure geospatial data availability and currentness, eliminate duplication in geospatial data production through increased coordination with producers and users, and transfer technologies to the private sector." The arguments in favor of cooperative activities include expected cost savings, data standardization, expanding data availability, and technology transfer. The partnerships in which the USGS is interested include conventional partnerships,

---

[69]White Paper, ibid.

[70]See http://www.usgs.gov/mou/ for the current MOUs.

innovative partnerships, framework partnerships, and Cooperative Research and Development Agreements (CRADAs).

In parallel with the development of the Archive and based on the recommendations of the National Performance Review, Executive Order 12906, *Coordinating Geographic Data Acquisition and Access: The National Spatial Data Infrastructure,* was signed by President Clinton on April 11, 1994. E.O. 12906's purpose was to have the executive branch develop a coordinated National Spatial Data Infrastructure (NSDI) to support public and private sector applications of geospatial data in transportation, community development, agriculture, emergency response, environmental management, and information technology.[71] It laid out the duties and responsibilities of the Federal Geographic Data Committee (FGDC) in coordinating the NSDI. The FGDC is chaired by the Interior Secretary or his/her designee, and all Executive branch departments and agencies with interests in the development of the NSDI are urged to provide a senior representative to the FGDC.[72] (To date, 17 federal agencies that make up the FGDC are developing the NSDI in cooperation with state, local, and tribal governments, the academic community, and the private sector.) The Executive Order states that the FGDC "shall seek to involve State, local, and tribal governments in the development and implementation of initiatives" described in the E.O., such as developing a National Geospatial Data Clearinghouse, standardization of geospatial data, establishing procedures for public access to geospatial data, and ensuring government agencies use the Clearinghouse before undertaking individual efforts to obtain geospatial data.[73] Furthermore, the E.O. specifies that within 9 months of the date of the Order, the Interior Secretary shall develop "...strategies for maximizing cooperative participatory efforts with State, local, and tribal governments, the private sector, and other nonfederal organizations to share costs and improve efficiencies of acquiring geospatial data..."[74] Exempt from E.O. 12905 compliance are the national security-related activities of DoD and Department of Energy (DoE), and intelligence activities as determined by the DCI.

This is an area that promises to grow in importance as technology develops and opportunities for collaborative activity between the U.S. government and private

---

[71]Executive Order 12906, *Coordinating Geographic Data Acquisition and Access: The National Spatial Data Infrastructure,* signed by President William J. Clinton on April 11, 1994.

[72]Ibid.

[73]Ibid.

[74]Ibid.

sector emerge,[75] and so could potentially offer a viable alternative programmatic approach to future interagency program concepts.

Alternative Approach 6: Joint Program Office

The last alternative approach is that of the joint program office. This approach is defined as a single, integrated program independent of, but responsive to, its parent organizations. The NPOESS program is a prominent example of this approach to interagency or joint acquisition and management. Other examples include the Global Positioning System (GPS), the Arsenal Ship, the Joint Strike Fighter (JSF), and the Discoverer II program.

The reader should review Appendixes A (NPOESS) and B (Arsenal Ship) for an examination of this alternative in greater detail. In addition, an August 2000 study conducted by three Defense Systems Management College (DSMC) Military Research Fellows on transatlantic armaments cooperation provides an excellent case study analysis that offers insights appropriate for interagency acquisition.[76]

DoD Regulation 5000.2-R defines a joint program as:

> Any acquisition system, subsystem, component, or technology program that involves a strategy that includes funding by more than on DoD Component during any phase of a system's life cycle shall be defined as a joint program. Joint programs shall be consolidated and collocated at the location of the lead component's program office, to the maximum extent practicable. This includes systems where one DoD Component may be acting as acquisition agent for another DoD Component by mutual agreement or where statute, DoD directive, or the USD (A&T) [Under Secretary of Defense (Acquisition and Technology)] or ASD(C3I) [Assistant Secretary of Defense (Command, Control, Communications, and Intelligence)] has designated a DoD organization to act as the lead....[77]

Within the DoD, the JROC for Acquisition Authority (ACAT) I programs or the Principal Staff Assistant (PSA) for ACAT IA programs review mission needs statements from the Services and operational requirements documents to

---

[75]Bruce Cahan, President, Urban Logic, Inc., *Financing the NSDI: Aligning Federal and Non-Federal Investments in Spatial Data, Decision Support and Information Resources, Executive Summary,* draft report for public comment, February 29, 2000, found at http://www.fgdc.gov/funding/urbanlogic_exsum.pdf.

[76]Lieutenant Colonel Richard C. Catington, USAF, Lieutenant Colonel Ole A. Knudson, USA, and Commander Joseph B. Yodzis, USN, *Transatlantic Armaments Cooperation: Report of the Military Research Fellows DSMC 1999-2000,* (Fort Belvoir, Virginia: Defense Systems Management College Press, August 2000). See Figure 4.1 in particular for a summary of the cases examined.

[77]Quoted in Eller, op. cit., pp. 1-2. DoD 5000.2-R was being revised as this document was being written.

determine whether potential exists for establishing a joint program.[78] The actual decision to establish a joint program is made by the Milestone Decision Authority (MDA), e.g., USD(A&T), based on the recommendations of the JROC for programs that will be reviewed by the Defense Acquisition Board (DAB), the recommendation of the functional PSA and ASD(C3I) for programs to be reviewed by the Major Automated Information System Review Council (MAISRC), or the recommendation of the DoD component (or designated representative) for all other programs.[79] Agreements between component MDAs provide the basis for the joint program, or they are directed top down from Congress or the USD(A&T). Mission needs statements are coordinated between the program partners, and programmatic milestones are established early in the schedule. Joint programs are managed through the lead agency or Service's acquisition hierarchy. Adjudicating among the different players in the joint program—not only the sponsoring agencies involved, but also Congress, the administration, and industry—can be difficult because of the differing requirements processes and the visibility of the program funding.

MOAs and MOUs form the basis for the joint program and provide the details of program organization and management prior to program development. As identified in the DSMC *Joint Program Management Handbook* (2nd edition, July 1996), key issues addressed in the MOA/MOU are:

- *Management*: determine the program manager's scope of authority; establish program selection criteria; define participating organizations and management organizational relationships
- *Requirements*: establish program requirements, process for validating changes; define who can create changes
- *Security*: determine degree of risk, what will be controlled, and how control will occur
- *Funding*: determine funding source(s), share (ratios, amounts), and agree to funds control measures
- *Contracting*: type of contract; whose rules govern the contract (lead/participating)
- *Conflict resolution device(s)*

---

[78]Eller, p. 2. An ACAT I program is a Major Defense Acquisition Program (MDAP) estimated to require more than $355 million in RDT&E or $2.135 billion in procurement, or those designated by USD(A&T) to be ACAT I. ACAT IA programs are Major Automated Information Systems (MAIS) estimated by the ASD(C3I) to require greater than $30 million per year program costs, or $120 million total program costs, or greater than $360 million in life cycle costs. All numbers are in FY 1996 constant dollars.

[79]Eller, pp. 2-3.

- *Integrated Product Teams (IPTs):* to cover requirements, logistics, cost/performance tradeoffs, interface/configuration control, test and evaluation (T&E)[80]

A formal review process exists (with appropriate documentation requirements) and milestones are established based on the categorization of the program, the requirements of the DAB, and the recommendations of the MDA. Appropriate test and evaluation organizations also play active roles in joint programs, and combined developmental test and operational test (DT/OT) approaches are encouraged as ways to achieve cost and schedule savings.[81] Establishing funding mechanisms and determining possible penalties for a participating agency's decision to reduce program funding or withdraw entirely are done by the lead agency or Service, reviewed by the JROC and DAB, and approved by USD(A&T). This is an important consideration for the program manager, given the extent to which he or she has to expend effort in maintaining stakeholder and agency funding support. Consequently, the program manager should understand the joint environment, the differing perspectives of the participating organizations, and how the users will utilize data or information from the system to accomplish their objectives.

Based on the literature review and interviews with a wide range of people familiar with joint programs, the most critical, difficult and time-consuming aspect of joint program management is probably the requirements process, particularly adjudicating among a diverse set of participating agencies each with their own requirements process or some with none at all. In contrast to many civil agencies, the DoD maintains a rigorous requirements process that begins with the Integrated Priority Lists (IPLs) originating with the CINCs based on their operational needs. The IPLs lead to the OSD and Service acquisition processes including Mission Needs Statements (MNSs) and to ORDs which establish objectives (most operationally meaningful, time critical, cost effective levels of performance), thresholds (minimum levels of performance necessary to meet user needs), and key performance parameters (those capabilities and characteristics so critical that their failure could cause program reassessment or cancellation).[82] A schedule of program milestones based on systems engineering phases provides key decision points for the life cycle of joint programs. While this schedule also applies to single agency (e.g., DoD) programs, what complicates the life cycle of joint programs is the necessity to coordinate differing processes and milestones among the participating agencies, whether all the

---

[80]Eller, pp. 12-13.

[81]Eller, pp. 15-16.

[82]Eller, pp. 33-35.

participating agencies adhere to the schedule—or more importantly, whether they have a rigorous process similar to DoD's that lends itself to making key "go/no go" program decisions at critical points in development and production.

Finally, while there is no single approach to joint program management, an important consideration is the negotiation of a charter between the participating organizations as a means to identify key factors and elements of the relationship, such as designating responsibilities and authorities for particular portions of the program. This charter would include a statement of common objectives, the program manager's role, funding and participation rules (including penalties for withdrawal from the program), joint program organization, staffing, and assignment of key leadership positions, methods for resolving conflicts among partner agencies, and other considerations such as political factors and changes in the threat.[83]

Evaluation Comparison and Insights

We will now compare the alternative approaches across element sets (e.g., acquisition complexity, requirements management, etc.) in Tables 3.4 through 3.11 below to compare and contrast approaches and identify potential broad insights for consideration by the Air Force and the NRO in collaborative activities they may undertake. There may be no "one right way" to approach interagency program acquisition and management, but insights based on historical experience are useful for capitalizing on potential opportunities and avoiding programmatic pitfalls prior to initiating an interagency program.

Acquisition Complexity

To repeat our definition of "acquisition complexity," this element denotes the degree of difficulty involved in acquiring a particular program or capability. It includes efforts to ensure that the program satisfies existing policy and objectives guidance and "front end" incentives to form the IPO.

In each case where a single lead agency acted to either execute the program, as the government representative in data or information buys from a variety of sources (e.g., NSLRSD), or to facilitate consolidation of common functions from a number of organizations involving significantly different institutional legacies (as in NIMA's case), interagency acquisition complexity was difficult but still easier than having two agencies designated equal leads and partners in interagency system acquisition ("equal" meaning equally shared responsibilities

[83]Eller, pp. 65-69.

from the top down, a potentially difficult approach managerially). Adopting streamlined acquisition approaches such as those characterizing ACTDs also encouraged the potential for faster fielding and strengthening of user or stakeholder support in the program. Furthermore, a system integrator approach might lend itself to the program benefiting from shared expertise and experience of the partner organizations as each partner or nation contributes a key part of the overall program.

On the other hand, while the system integrator approach would appear attractive, especially as a single overseer to handle complex processes and relationships, its authority could be weakened by the potentially opposing agendas and objectives inherent in its partners' internal political decisionmaking processes. This is particularly true for international programs such as the International Space Station where individual national interests may conflict with common program interests or where conflict resolution mechanisms or major program changes require unanimous decisions. An independent agent would face potential problems and issues of dealing with complex and dynamic technologies such as imagery and geospatial information technologies if it does not possess the institutional legacy knowledge or experience to carry out large scale, complex systems acquisition. Furthermore, consolidations of multiple, disparate programs of these kinds can make acquisition integration very difficult and potentially not very cost effective.

Table 3.4 compares the various approaches within the element of acquisition complexity.

Program Management

"Program management" as we defined it earlier refers to the organization, structure, and approach taken within a program to accomplish objectives.

Looking across alternative approaches, we see that strong program management depends on ensuring that program management objectives are consistent with common visions, goals, and policies developed by the parent organizations. Furthermore, support from the senior leadership of participating agencies is required to enable the program to achieve its objectives. These observations are true for almost all programs; their applicability here lies in the difficulty of being able to resolve disparate visions, goals, and objectives from the parent agencies or participants sufficiently to identify and achieve common goals for the program. This resolution needs to occur as early as possible to enable the lead agency in each approach to develop and implement an MOU or MOA that can be agreed to and implemented by the participants.

Table 3.4 Comparison of Approaches: Acquisition Complexity

APPROACH	STRENGTH	WEAKNESS	OPPORTUNITY	THREAT
EXECUTING AGENT	Most streamlined strategy using either traditional acquisition approaches or innovative concepts	Differing or conflicting organizational objectives and expectations may increase acquisition complexity at onset	Achieve program goals, field new capability in shortest time	Political, funding, technological, or programmatic risk can potentially be high
SYSTEM INTEGRATOR	Single acquisition organization to handle complexity of processes, relationships	Partner differences in objectives, processes may increase overall program complexity	Take advantage of individual partner expertise, experience	Loss of acquisition of key program elements if partner leaves program
INDEPENDENT AGENT	Single functional focus should facilitate acquisition approach	Insufficient experience at large complex system acquisition	Streamline acquisition approach, execution	Consolidation of disparate programs from multiple agencies can make acquisition integration very difficult
CONFEDERA-TION	Innovative partnership approach to acquisition, e.g., grants, foundations	Emphasis on coordination among diverse agencies	Consolidate multiple programs, products into one area	Potentially conflicting schedules, quality of products
JOINT PROGRAM OFFICE	Single agency assigned responsibility, authority for acquisition; single acquisition process determined at onset	Acquisition processes may not be compatible among parent agencies; schedule may be too ambitious	Streamline acquisition authority, responsibility; use of innovative approaches	Departure from, circumvention of traditional acquisition processes threaten organizational interests

In the Executing Agent approach, limited program objectives, such as a technology demonstration, should enable a program management structure that facilitates streamlined management and implementation and accomplishment of objectives in a short period of time. Conversely, the System Integrator and Confederation approaches will likely have longer duration and potentially greater organizational and program management complexity. Consequently, a greater premium should be placed on clear statements of intent, objectives, and metrics for success early in the development of the MOU/MOA and implementation plan. The System Integrator should be structured to maximize integration of key program elements and to enable effective government program oversight, but the difficulty in carrying this out among the range of participating agencies may contribute to program schedule slips and cost overruns. The Independent Agent approach offers an opportunity to consolidate similar functions inherited from parent organizations into one organization and to develop an integrated approach to management, acquisition, and program implementation. However, the agency's leadership and effectiveness in providing functional leadership and meeting stakeholder needs may be diminished if internal organizational cohesion is not well established and maintained. Finally, the Joint Program Office approach offers an opportunity within a structured management and organizational process to effectively implement common goals of a dual-use technology program such as GPS or NPOESS. But its effectiveness depends on partner participation and agreement on command relationships, authorities, and responsibilities that are established early in the program formulation and implemented effectively over time. All approaches need to have a management approach sufficiently flexible to respond to changes in threat, mission, vision, or other top level guidance. The management approach should identify metrics for program success, including conditions that may mean program termination when program goals are achieved or the capability or technology is effectively integrated into the users' operations.

Table 3.5 compares IPO approaches within the element of program management.

Program Control

"Program control" is the ability to monitor and influence the operation of a program by the responsible individual, i.e., a program manager; this is also called "span of control." This element focuses on the responsibilities and authorities of the program management team, the functional assignments agreed to among the participating agencies, and the communication flows within the program and to the stakeholders and users. Enabling the program manager and his/her team to effectively carry out the goals of the interagency program through the use of appropriate management tools is critical to program success. Trust and accountability are key factors in program control in all approaches. Trust among the partners and within the project team is crucial, as many of the approaches emphasize a streamlined approach to program span of control, i.e., potentially fewer reporting and documentation requirements in order to achieve shorter program schedules. If the mechanism for program control is not structured appropriately and supported by senior leadership in the parent agencies, potential opportunistic behavior could result by some of the participants, leading to schedule slips and cost overruns. Accountability is equally important for all approaches, especially in terms of identifying whether authority and responsibility are collocated to encourage accountability within the program. Furthermore, existing and potential contractor relationships will affect

Table 3.5 Comparison of Approaches: Program Management

APPROACH	STRENGTH	WEAKNESS	OPPORTUNITY	THREAT
EXECUTING AGENT	Structured program management; support from senior leadership, user communities	Coordinating multiple agency processes or ensuring no detrimental effect on program	Potential to streamline command relationships, chain of command to accomplish objectives	May run greater risk of turf battles with other agencies than traditional programs
SYSTEM INTEGRATOR	Structured to maximize integration of key program elements, maintain government program oversight	Extremely difficult, depending on number of participating agencies or partners	Undertake complex program using streamlined approach to integrating diverse elements	Schedule slips, cost overruns hurt overall progress of program
INDEPENDENT AGENT	Single functional focus should enable streamlined approach to program management	Ability of existing program management to deal with changes to threat, mission, vision, etc.	Opportunity to streamline program management	Potentially diminishing program leadership, effectiveness with other agencies performing similar functions
CONFEDERA-TION	Confederation approach for coordination of effort towards single objective, e.g., data buy; premium on establishing specific MOUs/MOAs early on	Size of effort may be too daunting for efficient program management	Reduce duplication of effort; consolidation of programs for specific objectives	Premium placed on coordination, establishing standards
JOINT PROGRAM OFFICE	Streamlined program management, supported by senior leadership in parent agencies	Potential disagreements over required resources, developmental tasks, optimistic designs	Go "outside the system" to accomplish program, acquisition goals	Lack of established multi-agency corporate support for program at early stages could be critical later

program control, particularly in their flexibility, the extent of certainty or uncertainty in the contracts, and in the contractual governance structure.

Table 3.6 provides a comparison of approaches for this element.

Requirements Management

"Requirements management" is defined as involving the adjudication, coordination, and implementation of a common requirements process for the program. Understanding each agency's institutional approach to generating and adjudicating among a diverse set of requirements will enable the IPO program

Table 3.6 Comparison of Approaches: Program Control

APPROACH	STRENGTH	WEAKNESS	OPPORTUNITY	THREAT
EXECUTING AGENT	Span of control is streamlined, with fewer reporting, documentation requirements; emphasis on mutual trust	Innovative approach may require alternative approaches to traditional chains of command	Exercise alternative approaches to program authorities, responsibilities	If trust not present, may lead to opportunistic behavior by participants at expense of achieving program goals and schedule
SYSTEM INTEGRATOR	Build on functional insight into program development	May be difficult given number of partners	Balance program control, oversight with encouraging innovation	System integrator may not have sufficient control over partner activities
INDEPENDENT AGENT	Span of control should be eased by functional, program consolidation	Too great a span of control encourages micro-management at cost of getting things done	Opportunity to exercise short span of control	Lack of delegation of authority, responsibility may slow progress of program
CONFEDERATION	Requires senior steering group, external corporate (agency) support	Efficient span of control may be extremely difficult to achieve	Take advantage of comparable efforts, build extensive outreach	May be too spread out to be effective
JOINT PROGRAM OFFICE	Emphasis on less detail of documents, more trust in contractor relationships	Lack of sufficient specificity could lead to misinterpretation over requirements satisfaction	Better working relationships, trust between program office and contractors	Potential misinterpretation of government, contractor motivations

management team to account for potential differences and impacts on program schedules. This is a particularly critical element for an interagency program, given the stakes involved and the importance in carrying this out to the satisfaction of the participating agencies. Requirements adjudication may require the greatest amount of attention of the program manager and his/her staff, not only at the beginning of a program, but also later to prevent "requirements creep."

Each of the approaches addresses requirements management slightly differently. In the Executing Agent approach as exemplified by ACTDs, determining military utility is the critical measure of program success and is identified through meeting user needs and increasing user familiarity through technology demonstrations in a realistic manner through operational exercises. Viewed as a demonstration and interim solution to meeting user needs prior to full scale acquisition and production, requirements creep is thus minimized, unlike possible outcomes in the other approaches, particularly the Independent Agent and Confederation approaches. Because the Independent Agent has functional responsibility and leadership, it may be in its best interest to encourage requirements expansion as a way to increase stakeholder support. But, without a viable requirements process in place, this may also increase the difficulty in adjudicating among an expanding set of requirements. The example applied in this approach was NIMA, which as a DoD Combat Support Agency employs the well established, rigorous DoD requirements process. In notable contrast, NASA is responsible for systems integration of the ISSP and while acting as the focal point for requirements, has to negotiate among potentially conflicting national interests of the participants. Here the MOU/MOA is the key document for establishing common requirements at the onset of the program and is crucial for ensuring specific objectives for the system contribution made by the participating nations are explicitly identified. Measures to protect against divergence of requirements over time based on changes in interests and priorities should be factored into the program management and implementation plans of all approaches at the onset.

A comparison of approaches is shown in Table 3.7 below.

Table 3.7 Comparison of Approaches: Requirements Management

APPROACH	STRENGTH	WEAKNESS	OPPORTUNITY	THREAT
EXECUTING AGENT	Emphasis on determining mission need/military utility	If new program or concept, may not have supporting doctrinal or requirements rationale in place yet	Link CONOPs and requirements	Requirements agencies may not buy it
SYSTEM INTEGRATOR	Consolidation of requirements in one point of contact	Diversity of partners means potential difficulty in adjudicating among conflicting requirements	Identify common goals that no single agency, government could accomplish on own	Divergence of requirements over time based on changes in interests, priorities
INDEPENDENT AGENT	Consolidates functional requirements in one place	Difficult to adjudicate among wide range of requirements	Conduct functional tradeoffs among air, sea, land, spacer-based capabilities to meet requirements	Lack of adjudication mechanism for requirements may stall program
CONFEDERA-TION	Alliance sets requirements, standards for community effort	Adjudicating among requirements, different standards	Consolidate all requirements, standards in common effort	Lack of support for program because of different objectives, goals
JOINT PROGRAM OFFICE	Established process to adjudicate among diverse set of requirements and users	Difficulty of adjudicating, justifying requirements among diverse communities	Focus discrete mission, system requirements	Potential mission requirements "creep"

Funding Stability

"Funding stability" is considered here as the process of maintaining funding support among the organizational partners over the lifetime of the program. Assessments of the effects of potential changes in requirements or participant priorities on funding stability are necessary at decision-making milestones in the program development. Furthermore, the partners must determine the ratio or proportionality of cost sharing when the MOU/MOA is developed, and identify penalties for reducing cost shares or withdrawing from the program in order to plan for possible contingencies.

Funding stability is crucial for all programs, whether they are interagency or not. The applicability of this requirement here lies in the approaches taken to ensuring funding stability in each IPO approach. In the Executing Agent approach, the objectives and schedule are limited, e.g., demonstration and/or fielding within three years, so funding schedules are short term in duration. Nevertheless, planning for transitioning the technology demonstration to full scale development, assuming military utility is proven, does account for life cycle funding requirements. In the System Integrator case, ensuring funding support also can help to ensure buy-in to the program. But as we have seen in the ISSP example, funding slips by one partner may have a critical and adverse impact on the success of the entire program. An Independent Agent may offer a way around this problem by consolidating funding within the organization; this encourages more effective tradeoffs among system and technology opportunities matched by available resources to meet user needs. A Confederation may offer alternative approaches to funding stability, such as public-private partnerships, grants, and cooperative agreements to encourage stable financial investment. The JPO should also encourage funding stability, but (like many others) will be subject to the parent agencies' budgets. In this particular case, also true with others, it is in the best interests of the JPO leadership to work with the appropriate congressional committees and staffs to ensure its interests and funding needs are effectively represented.

Table 3.8 below compares the five approaches within the element of funding stability.

Table 3.8 Comparison of Approaches: Funding Stability

APPROACH	STRENGTH	WEAKNESS	OPPORTUNITY	THREAT
EXECUTING AGENT	Limited program objectives and shortened schedules could lessen long term funding burden	Ability to ensure funding in transition from demonstration to full program	Accomplish specific objectives with limited funding using innovative approaches; transition planning for full scale development	Bypassing traditional authorities within hierarchy may lead to funding cuts
SYSTEM INTEGRATOR	Partners' funding contributions enhances program buy-in	Complexity in funding from multiple sources contributes to program instability	Maximize opportunities to fulfill goals, raise program visibility with Congress	Potential schedule uncertainties complicate funding schedules; increased oversight by Congress
INDEPENDENT AGENT	Facilitates funding focus on single program or functional area	Potential conflicting funding sources from other agencies with legacy systems, capabilities	Split funding (e.g., between military and intelligence community) offers potential funding stability	Severe implications if one parent agency cuts its contribution
CONFEDERA-TION	Mix of grants and cooperative agreements to construct efficient program, encourage stable financial investment	Need for consistency in funding approaches, or ability to cope with funding uncertainties	Potential opportunity to fund through consortia of regional parties, industry, and/or interest groups	If not well established early in program, may not lead to robust effort; implications for future opportunities for alliances of this type
JOINT PROGRAM OFFICE	Drawing from multiple sources reduces funding vulnerabilities	Dependence on multiple sources results in increasing oversight complexity; insufficient resources; competition with other priorities	Balance sources of budget; emphasis on affordability; build broader support for program	Potential cuts in parent organization funding; competition for scarce resources; place in priority chain; programmatic under-funding overall

Customer Responsiveness

"Customer responsiveness" refers to the program's relationship to its users and stakeholders. As noted earlier, ensuring customer and stakeholder support for the program will consume much of the program management's time and energy. This activity also serves as an important indicator of program success. While these statements apply to programs in general, applying them to interagency programs greatly complicates both the time and energy required of the program leadership and can have a tremendous effect on program success, particularly if not effectively carried out.

Each IPO approach has to identify its stakeholders in a variety of communities (e.g., the Administration, the Congress, the public sector, commercial industry, and international); determine the extent of oversight, the oversight process, and its effect on the program; and determine how much time will be necessary for the program leadership, especially the PM, to devote to ensuring program success. In the Executing Agent example, the program structure itself encourages customer responsiveness by having key organizations and interests actively represented in the program development process. The System Integrator offers a single agency point of contact for issues, concerns, and influence, albeit partner participation in ensuring stakeholder support back home is crucial. By consolidating functions within one organization, the Independent Agent inherits the stakeholder support requirements from its legacy organizations and, as in NIMA's case, may have to respond to potentially conflicting demands from stakeholders in both the defense and national intelligence communities. This path can be a difficult one to weave unless common objectives and program goals are agreed to by the stakeholders, including congressional oversight committees. The JPO may also have the same problem, again depending on its partner agencies, especially whether they come from the DoD community exclusively or whether they also include civil agencies as in the NPOESS case.

In all cases, customer responsiveness is an area that is critical to program success and will likely require far more time and effort than originally envisioned. Shown below is Table 3.9, again a comparison of approaches regarding customer responsiveness.

Cultural Alignment

Organizational culture can influence organizational capabilities, assigned program personnel and expertise, and program structure and process. Here we define "cultural alignment" as the interaction of and implications for the program of the diverse organizational cultures inherited from the parent or partner organizations. Below we consider the implications of cultural alignment on each of the IPO approaches, including the strength and weakness of the approach in response to this element, the opportunity posed by cultural alignment to the approach, and the potential threat. In the Executing Agent approach, the effect of cultural alignment may be minimal depending on the agency carrying out the program, or in sharp contrast, it could require radical

Table 3.9 Comparison of Approaches: Customer Responsiveness

APPROACH	STRENGTH	WEAKNESS	OPPORTUNITY	THREAT
EXECUTING AGENT	Direct correlation between program, client base; fielding of prototype could encourage greater system acquisition support	Maintaining support for program in near term and for transition to full program	Improve support to warfighter in specific areas more easily definable, amenable to performance metrics	User and developer communication often poor
SYSTEM INTEGRATOR	Customers have single point of contact for issues, concerns, influence	Customers can influence partners to gain indirect advantage	Maximize potential customer base	Balancing diversity of customers may lead to dissatisfaction with service provided
INDEPENDENT AGENT	Single point of contact to meet customer needs	Potentially single point failure in responsiveness	Consolidate multiple programs, establish single point of contact for customers	Trying to satisfy myriad of customers with different needs
CONFEDERA-TION	Consolidation of user communities into alliance network	Cooperation and collaboration are critical	Establish working relationships with wide variety of user communities	User and congressional oversight may be difficult
JOINT PROGRAM OFFICE	Involving users in program leads to buy-in	Misunderstandings over program goals, objectives	Build integrated CONOPS, enhance operational effectiveness	Lack of support from parent agencies, Congress could kill program

organizational change. For the System Integrator approach, cultural alignment could potentially be not much of a problem if integrating system elements within the total program is a straightforward process. On the other hand, as the ISSP demonstrates, different cultural values and perspectives, found in government-industry relationships for example, could have a dramatic effect on the ability of the system integrator to effectively oversee the program. Again, a premium is placed on the initial MOU/MOA in order to minimize potential disruptions or outcomes of different cultural alignments on the program.

The Independent Agent approach offers an opportunity to build on organizational cultural legacies and fashion a new overarching organizational culture. Its weakness, however, is the potential inability of the organization to overcome its legacy cultures that can inhibit organizational cohesion and cultural alignment. NIMA is an example of an organization built on the imposed merger of two distinct cultures (military, intelligence community) that has not yet developed a distinct culture of its own. It remains to be seen whether NIMA can surmount its cultural legacies to form a new geospatial information-based culture, one that is aligned with its mission, vision, and core competencies.

The strength of the Confederation approach with regard to cultural alignment lies in the importance of and motivation for the alliance to overcome challenges to confederation cohesion as the members of the confederation pursue shared program goals. It also offers an opportunity to minimize the cultural effects of multi-agency participation by specifically identifying the roles and responsibilities of confederation members early on in the initiation of the program. The challenge to this approach, though, lies in potential interoperability problems stemming from different cultural approaches to the acquisition process and their effect on confederation cohesion and accomplishment of program objectives.

Finally, the Joint Program Office, like the Independent Agent, offers the opportunity to surmount legacy organizational cultures by collocating personnel with different backgrounds who are assigned as staff to the program. In DoD joint programs, this can mean officers and enlisted personnel from each of the Services who bring distinctly different operational experiences and cultural backgrounds to the program, yet are united by familiar DoD acquisition processes and procedures. Interagency programs between the DoD and civil agencies, such as NOAA, may be much more affected by cultural alignment because of the absence of similar requirements and acquisition processes, compounded by different staff management approaches. An example of this effect lies in the institutional memory possessed by long term civil program staff versus military staff who rotate every 18 months to two years. A potential weakness may be a perception that JPO program management "favors" one organizational culture over another, e.g., "we do this the Air Force way" rather than the "joint way." This perception could occur perhaps as a result of one Service providing the bulk of the program personnel, but can potentially be overcome if treated in the MOU/MOA through the specific assignment of parent agency responsibilities and staff.

Table 3.10 illustrates a comparison among approaches in cultural alignment.

Staffing

"Staffing" is concerned with the staffing process of the program and the ability to attract qualified personnel to work in the program. How each approach develops this ability depends on the inherent qualities and characteristics of the program, e.g., innovation, opportunities for further career progression based on program experience, etc. As shown in the ACTD example, the Executing Agent approach may offer innovation, a streamlined acquisition process, and the

Table 3.10 Comparison of Approaches: Cultural Alignment

APPROACH	STRENGTH	WEAKNESS	OPPORTUNITY	THREAT
EXECUTING AGENT	May be minimal depending on agency	Requires radical change	May be opportunity to address cultural integration through operational fielding of exploratory system/capability	Unfamiliarity of user with new capability
SYSTEM INTEGRATOR	Not as much of problem if partners contributing program elements	Potential cultural differences in approaches to programs, relationships between government and industry	Expand cross-agency, cross-cultural understanding, willingness to undertake future shared efforts	Problems may contribute to mistrust of system integrator, unwillingness to conduct future cooperative programs
INDEPENDENT AGENT	Potential to develop overarching culture with clear focus (one function)	Legacy cultures may inhibit cultural alignment, organizational cohesion	Single functional focus facilitates cultural cohesion/alignment with vision, mission, core competencies	Culture may not align with vision, mission, core competencies
CONFEDERATION	Alliance objective may outweigh cultural aspects	Cultural differences may outweigh shared alliance objectives	Minimize cultural effects of multi-agency participation through specificity of roles, responsibilities	Potential interoperability issues
JOINT PROGRAM OFFICE	Collocation of staff with different cultural legacies	Perception that JPO favors one culture at expense of integrated culture	Enhance working level relationships among staff from different backgrounds, parent organizations	Effect of cultural legacies on staff motivations, support for program

potential opportunity for an individual to see a program or system fielded during that individual's involvement in the program. The System Integrator and Confederation approaches promote staffing from personnel from the lead agency, while the Independent Agent and JPO approaches build on staff expertise and experience inherited from the parent agencies. Programs using each of these approaches will inevitably compete for skilled people with either larger agency programs, longer term ones, or the private sector, so the imperative rests with program management to identify incentives to attract the best qualified people for their programs and to minimize potential disincentives for interested applicants. A comparison of approaches with respect to staffing is shown in Table 3.11.

Next, Chapter 4 will address conclusions and insights regarding interagency program concepts for Air Force-NRO integration activities.

Table 3.11 Comparison of Approaches: Staffing

APPROACH	STRENGTH	WEAKNESS	OPPORTUNITY	THREAT
EXECUTING AGENT	Innovative program will attract qualified staff	Competition with other longer term programs for staff	Integrate nontraditional thinking from commercial, other sectors into program	May be too innovative to be accepted by traditional hierarchy outside program
SYSTEM INTEGRATOR	IPO staffing from lead country, agency	Liaisons from other countries, agencies	Credibility of system integrator to gain best qualified staff	Agencies may not send most experienced staff to participate
INDEPENDENT AGENT	Identify incentives, new approaches to hire most qualified people	Not competing with private sector very successfully	Exploit unique capabilities to get most experienced people	Better career opportunities at parent organizations; legacy personnel systems
CONFEDERA-TION	Lead agency provides management staff	Dependent on participating agencies for external staff, liaison support	Minimize lead agency staff requirements?	Alliance structure may preclude innovative approaches to gaining best staff
JOINT PROGRAM OFFICE	Successfully drawn from parent agencies; in some cases, hand-picked	Military staff turnover rates result in loss of corporate knowledge	Expand credibility of JPO as good career opportunity for individuals	Parent agencies do not send their best people to JPO

4. Conclusions and Insights

This study was conducted to assist the ANIPG in identifying potential concepts for integration by analyzing interagency program management approaches and implementation. Six alternative approaches to interagency program management were identified, and five were addressed in depth by using case study examples to illustrate various aspects of the approaches. This enabled us to identify appropriate insights for potential Air Force-NRO integration activities, including decisions, actions, and mechanisms that worked and those that did not.

The motivation and interest in engaging in cooperative or integrative activities and in interagency programs are occurring because of increasing overseas commitments requiring intelligence and space system support, funding constraints and increased congressional scrutiny of military and intelligence programs, and pressures to eliminate program redundancies and inefficiencies. While our focus here is on the Air Force and the NRO, other agencies are facing similar pressures to conduct joint or interagency programs. The NPOESS program is a prominent example of an interagency program conducted between NOAA, DoD, and NASA. It is likely that such cooperative programs and activities will be encouraged in the future, assuming sufficiently common organizational interests and requirements.

As we saw in Figure 2.1, opportunities for integrative activities may occur at many points in time for a variety of objectives. Commonality of interests may occur in the development of CONOPS, in determining required capabilities to perform certain operational tasks and achieve specific national security or military objectives, or in identifying an opportunity for an interagency program to provide specific capabilities to meet user needs. Figure 2.1 also represents an iterative process, for stakeholders and users will provide feedback on the successful integration or application of a program to meet their needs.

Given the level of interest in interagency programs, what is the "best" approach to implementing them? No single approach stands out as clearly the best way; rather, each potential program should be tailored to fit the organizational and programmatic objectives and to meet the needs of the stakeholders and users. This is true of all acquisition programs; however, the distinction for interagency programs occurs when considering the increasing complexity involved when integrating two (or more) different organizational processes, interests, and acquisition approaches to key program elements such as mission requirements, funding, policy and regulatory requirements, and oversight. These differences need to be understood at the initiation of discussions among agencies and other interested parties and at the negotiation of an MOU or MOA establishing the program. Attempting to solve them after the program has begun offers the potential for misunderstanding between the partners, schedule slips, increased congressional scrutiny, and funding instability.

Insights and Observations

Numerous insights and observations stand out as being important or even essential to IPO success, and are consistent with both the questions raised in Chapter 2 and as evidenced by the case studies and background research conducted for this study. They include the following, again divided along the lines of the SWOT elements. Since these elements are interrelated, any consideration of a potential integration concept involving the use of an IPO should not address them in isolation, but should weave them together into a coherent program strategy that makes sense for the particular concept in mind. Again, while this holds true for most acquisition programs, the multi-agency nature of an IPO will complicate every element of the program strategy, but this should not be considered an insurmountable problem.

Acquisition Complexity

The importance and criticality of support from the leadership of each participating organization cannot be underestimated. Each partner agency's management must understand and accept the organizational agreements negotiated among the partners, for their support is integral to program success. When it comes to determining specific responsibilities, roles, and responsibilities for program activities at the parent organization level, high level organizational leadership support is also necessary for ensuring appropriate cooperation at lower levels within the parent organizations. This becomes important particularly during budgetary reviews for ensuring adequate funding stability and continuity.

Program goals and objectives should be consistent with higher level policy guidance, including national space policy and national security policy, defense planning guidance, relevant intelligence policies, and legal and regulatory agreements such as treaties, where appropriate. Of particular interest will be policies or regulations that guide or bind one partner organization but not the other. Security and interoperability considerations should be factored into the planning throughout, and potential concerns resolved. Linked to security considerations are jointly agreed-to mechanisms for maintaining IPO information infrastructure assurance against common threats. A survey of related or similar activities underway in other organizations should be conducted early on to identify unique applications and potential opportunities for further collaboration. For example, identifying programs or technology demonstration efforts underway in civil space agencies such as NASA may lead to useful to collaboration in solving particularly difficult or challenging technical problems. This has the added benefit of further expanding program support from other agencies.

Program Management

Following agreement on common program goals and agency interests, addressing program management aspects, specifically the MOU/MOA, is crucial to establishing the scope and organizational structure of the IPO. The MOU/MOA needs to be sufficiently robust to ensure parent agency support and to assign specific roles and responsibilities among participating agencies, yet flexible enough to respond to potential changes in policy and planning guidance, the threat, or other high-level factors. Also key is the development of an overall program strategy that includes implementation and funding strategies running the lifetime of the program, including termination, is consistent with national policy and guidance, agency goals and objectives, and regulations, and is executable at critical program milestones. The program funding strategy should include cost sharing arrangements among the parent agencies, and include possible penalties for withdrawal from the agreed-to arrangement (again, to encourage and facilitate overall funding stability).

Program planning and management also need to recognize and account for the potential challenges posed by differing parent agency planning and budgetary cycles and the increased burden in time and manpower placed on IPO leadership to deal with maintaining funding stability and stakeholder support. Furthermore, requirements adjudication will be a key issue to which the IPO program leadership need to devote significant attention and resources.

Program Control

Program control addresses the flow of information and communication through the integrated organizational chain of command, and is influenced by organizational structure and the program management's "span of control." The approach to program control also can influence or hinder the collocation of authority and responsibility to encourage accountability within the program. The program strategy discussed earlier needs to ensure that the program manager and his/her senior team have unimpeded access to the information they need to execute the program successfully. Complicating program control will be changes in laws and regulatory policies regarding privacy and confidentiality requirements, liability, and national security requirements, and the overall growth in information technologies and access to information via the Internet.

Organizationally, a strong, decisive executive council is needed to keep the program on track and to engage at senior levels of the parent agencies and elsewhere as appropriate to deflect potential problems or to identify particular organizational perspectives which may influence the program. Existing relationships among team members are an asset for the program management and should be encouraged whenever possible, especially at lower levels within the organization where the "real work" gets done. Clear indications of responsibility for program reviews and for "signing off" procedures to move to the next program milestone should be identified early. Furthermore, mechanisms for conflict resolution (i.e., disagreements among parent agencies or within the staff that are adversely affecting the program) should be available to program management, the earlier the better to minimize potential stress on program execution.

Requirements Management

This element may be the most difficult and time-consuming part of an IPO and should be managed effectively to minimize the natural tendency to have "requirements creep" in the program. "Requirements creep" refers to the tendency to add on additional requirements or "nice-to-haves" to a program as the program is underway. This can result in a perception of "gold plating" which will invariably invite increased administration attention and congressional scrutiny. Effective requirements management starts by understanding how each member organization of the IPO identifies user requirements, what metrics each organization employs to measure requirements success, and what process the provider organization uses to reach out to its customers. In DoD the process is well established, and provides rigorous traceability from requirements to capabilities. Similar processes do not exist in civil agencies, or if they do, they are much less rigorous or are driven by scientific processes that emphasize data collection and exploration rather than specific needs. Given the disparity among requirements processes, it is very important to identify and designate a preferred requirements process for the program in the starting MOU/MOA. Furthermore, this preferred process should include a requirements adjudication mechanism to minimize or discourage requirements creep by the partner agencies. Recognition also needs to be made of the greater than expected amount of time necessary for the program staff to deal with this issue that could have a direct bearing on program schedule and cost.

Funding Stability

Funding stability will be critical to the success of the IPO, therefore, mechanisms must be established in the early program planning stage to determine participating agency goals and interests, funding processes and schedules, and cost sharing arrangements before the program becomes a formal reality. The nature and type of a funding strategy that contributes to program stability will depend on whether the IPO is a technology demonstration program that will be quickly transitioned to a fully fielded capability, or a more traditional acquisition program for developing and procuring a large number of operational systems. This funding strategy should encompass the lifetime of the program and be agreed to by the participating organizations. Particularly important are penalties imposed on partner organizations for withdrawing from the program at unexpected times that would adversely affect program success, and at minimum, consideration of contingency funding sources and plans should that withdrawal occur regardless. As noted earlier, this action taken by a partner organization with far more resources than another partner can doom a program because of the inability of the other partner to compensate. The management team needs to take all considerations into account and make the program executive council aware of potential problems or concerns in time to influence or deter them.

Customer Responsiveness

Ensuring stakeholder support is another critical element to IPO success, and is complicated by the multiplicity of stakeholders influencing an IPO, some with conflicting goals and objectives. A leading example of this is the complication of additional congressional oversight committees and staffs that occur when considering IPOs that cut across civil and national security sectors. One approach to raising the level of customer awareness and support and to ensuring program awareness of differing agency perspectives is to involve them in the program from the earliest planning phases. This is a key part of the ACTD process, an example of the Executing Agent approach. As noted earlier, ACTDs represent opportunities to demonstrate advanced technologies to military forces in the field, thus encouraging experimentation prior to full scale acquisition and development. By involving operational commands in the ACTD process, and making military utility a key benchmark for ACTD success, this increases customer familiarity with and support for the program. The IPO should have military or overall national security utility as an essential element of measuring program success in order to ensure stakeholder support. But the necessity of dealing with a wider range of customers by an IPO will complicate the ability of the IPO leadership to address this easily.

Another aspect to customer responsiveness is the necessity to keep external interested parties such as other parts of the administration and Congress informed and aware not only of program successes, but also of potential or impending problems and concerns. While no one likes to get bad news, keeping the senior leadership routinely informed can help to minimize potential surprises later. The executive council can be used to help the program management in this regard. The requirement for routine communication with all IPO participants by the program manager and deputy program manager may influence the kinds of skills and experience needed for these positions, more so than in a traditional acquisition program. For example, individuals with direct experience and knowledge of acquisition and program processes in multiple agencies, especially those in different sectors (i.e., military, intelligence, civil), should be an asset.

Cultural Alignment

Organizational culture can influence both operational capabilities and organizational structure and process. By its nature an IPO will have multiple cultures represented in its organization and personnel. Some cultures may be similar insofar as they have similar institutional objectives and experiences. Others may be radically different, such as IPOs formed from military or intelligence organizations and civil or science-oriented organizations. Organizational culture may also influence program control in terms of information flows and command hierarchies. Recognizing the effect of these different cultures on IPO organizational structure, management, and execution is crucial in order to understand and deal with internal bureaucratic behavior and with external pressures coming from not only the parent organizations but also other interested parties. Whether it is necessary to develop an overarching IPO culture may depend on the specific situation and duration of the program. In the Independent Agent approach, an overarching culture is required as a way to enable the staff to see how they as individuals and their position within the organization contribute to overall organizational success. Clear statements of IPO goals and expectations are important to ensure the staff understands the criteria for mission and program success.

Staffing

Last, but not least, is staffing. Once the IPO leadership and parent agencies have identified goals and management objectives for the program, they need to carefully think through how they will obtain the most qualified and experienced people to support the IPO. If the IPO is military in nature, the IPO leadership will have to account for a fair amount of staff turnover at routine intervals. Turnover has a direct bearing on institutional memory, but an IPO can benefit by having staff from one of its partner agencies come from a community where staff longevity is routine. The NPOESS program is an example of this situation where institutional memory resides more in the civil employees and contractors than in the DoD side. Requirements for staffing success will include whether the IPO management identifies incentives to encourage new staff and retain existing staff, and if contingency plans are developed to account for the loss or transition of key staff members to other positions outside the IPO. Ideally, staff should view the IPO as an exciting place to *be*, but also as an exciting place to be *from*. Additional factors to be considered include IPO-required training and education, and post-IPO promotion opportunities.

As we have emphasized in this report, integration activities between the Air Force and the NRO can occur at various points in time and in various venues. Cooperation can occur in the development of CONOPS, in determining commonly required operational capabilities, and through the specific mechanism of an interagency program office. As noted earlier, integration and joint programs can potentially offer the benefits of providing joint combat and operational capabilities, improving interoperability among military and intelligence components and providers, reducing development and production costs, meeting multi-user needs, and reducing logistical requirements through standardization. As reflective of the Air Force's and NRO's interest in pursuing cooperative activities that both support warfighter and national decision-maker as well as the goals of aerospace integration, the IPO offers an approach to achieving those goals. Insights and observations derived from prior historical examples, whether successful or not, can help to alert those considering establishing IPOs to the pitfalls and benefits of this approach in comparison to other possible approaches. Finally, integrating key elements of an IPO such as those used in this report can help to achieve an integrated program strategy that accounts for programmatic requirements and user needs, deals successfully with unforeseen events, and achieves national policy goals and objectives. That is the minimum the American taxpayer and warfighter should expect.

Appendixes

Appendix A. NPOESS Case Study

Overview

One of the most ubiquitous applications of earth-orbiting, artificial satellites in the last 30 years has been monitoring planetary weather. Civil and military leaders in the United States and other countries have dedicated significant resources to deploy and operate expensive satellites that provide near real-time, highly accurate meteorological data to a large community of diverse users. The Department of Defense called their program the Defense Meteorological Support Program (DMSP), which has been in operation since the early 1960s. NOAA[84] has operated a very similar satellite program since 1960, which eventually became POES. The convergence of the two programs was initiated in 1994, under the new title of NPOESS.

The formation and early operation of NPOESS can serve as a valuable case study for other programs considering some form of joint program, especially within the American federal bureaucracy, and it is for this reason we believe it most closely fits within the Joint/Integrated Program Office alternative described in the main body of this report. The NOAA and DoD programs reflected significantly different priorities consistent with their overall missions, even as they operated similar spacecraft that carried out similar tasks. Before the creation of NPOESS, on eight occasions since 1972, officials studied the possibility of converging DMSP and POES, since it appeared to many that there might be unnecessary redundancy at the American taxpayers' annual expense and potential opportunities for greater efficiencies. The two programs did cooperate on a limited basis, yet convergence never occurred because of fundamental differences in DoD and NOAA requirements. The creation of NPOESS is a study on how two programs finally managed to overcome cultural and policy differences, while also meeting all mission needs.

---

[84]NOAA is a civil agency in the Department of Commerce (DoC). Within NOAA resides NESDIS, the National Environmental Satellite, Data, and Information Service, whose mission is to provide and ensure timely access to global environmental data from satellites and other sources to promote, protect, and enhance the nation's economy, security, environment, and quality of life. NESDIS manages the nation's operational environmental satellites, provides data and information services, and conducts related research.

DMSP

Military weather reconnaissance began in earnest in the early 1960s, to determine cloud cover over land areas targeted by aerial mapping surveys and top secret imaging satellites. The first officially designated DMSP satellite was launched in 1965. In the first few years, nearly a dozen of these short-lived satellites reached orbit. Later DMSP spacecraft were significantly improved, upgrading to electro-optical, then multi-spectral sensors and high-rate transmission of imaging data to ground stations. To date, the program has successfully deployed almost 40 satellites.

The standard operating procedure for DMSP was to maintain two spacecraft in near-polar, circular, sun-synchronous orbits at all times, primarily in support of national security requirements, but also to provide civil and military leaders with near real-time, global weather information. The satellite command and control center resided at Offutt AFB, Nebraska, in the 6th Space Operations Squadron (6SOPS), with a backup center at 6th SOPS Detachment 1 located near Fairchild AFB, Washington. Command authority for the program rested with 50th Space Wing at Schriever AFB, Colorado, an element of 14th Air Force (a component of Air Force Space Command). DMSP also took advantage of the extensive Air Force Satellite Control Network (AFSCN), which operates ground stations in many locations around the world. From top to bottom, DMSP was a fundamentally military program with very specific mission requirements that did not disappear with the end of the Cold War.

POES

The United States began investigating the feasibility of civil earth observation satellites in the early 1960s, inspired after NASA astronauts returned with valuable pictures of earth. Early programs such as Tiros and Nimbus paved the way for the Landsat remote-sensing program and several weather-monitoring spacecraft programs. NOAA became the focal point for the two key civil programs, the Geostationary Operational Environmental Satellite (GOES) and POES. NOAA operates a command and control center in Suitland, Maryland, with command relay and data acquisition facilities at Wallops Island, Virginia and Fairbanks, Alaska. To this day, GOES spacecraft provide continuous, wide-angle weather imaging from geostationary orbit. The POES spacecraft, operating much closer to earth in circular, near-polar, sun-synchronous orbits, gather more precise data for near real-time applications.

The first U.S. civil satellites were launched in the early 1960s under the TIROS, TOS, and ITOS programs, among others. These satellites, like their early military counterparts, were short-lived and crude compared to modern platforms. The first POES[85] satellite was launched in 1970 and until 1995, a total of 16 spacecraft have been built for civilian weather monitoring. Ground stations in 120 nations receive POES weather data, while thousands of schools, civil organizations, and private individuals also have access to NOAA imagery. POES has been a fundamentally civilian program serving public interests with a continuing mission to provide weather information to as many users as possible.

Convergence of DMSP and POES

By the early 1990s, it became clear that the convergence of POES and DMSP was not only possible, but also necessary in a post Cold War environment of reinventing government and tightening budgets. Vice President Gore sponsored a National Performance Review (NPR) concurrent with congressional committees drafting the Government Reinvention Act (H.R.3400), both authorizing and expediting the merger. On 5 May 1994, the White House issued Presidential Decision Directive NSTC-2, directing the convergence of POES and DMSP into NPOESS, which was expected to satisfy civilian and national security operational requirements. Also, NASA, in conjunction with its Earth Observing System (EOS) program, would offer new remote sensing and spacecraft technologies to improve the future NPOESS spacecraft. Finally, the President also directed DOD, DOC, and NASA to establish an IPO to manage the converged system.

By March 2000, the NPOESS IPO completed many of the primary goals of system convergence. The IPO was established on 1 October 1994 and presently operates from offices in Silver Spring, Maryland. The primary command and control facilities and data distribution center for POES and DMSP have been centralized in Suitland, Maryland, while the backup facilities for DMSP became operational at Schriever AFB near Colorado Springs, Colorado. Operational space assets include two primary DMSP and two primary POES spacecraft, with a number of on-orbit backups of various ages and system health. NPOESS will continue to have access to AFSCN resources as DMSP did before, adding to the same ground stations used previously by POES.

In 2000, convergence of the DMSP and POES programs is not complete. It is more precise to say that the NPOESS IPO represents the sole agency that will develop, acquire and operate future NPOESS spacecraft that have yet to be constructed and will eventually replace all older DMSP and POES spacecraft.

---

[85]POES spacecraft are more commonly called NOAA satellites, a designation that sometimes causes confusion. Also, NOAA (POES) satellites have two designators, such as NOAA-14 (NOAA J). This is because NOAA assigns a letter to the satellite before it is launched, and a number once it has achieved orbit.

Currently, the IPO has responsibility (satellite control authority) only over the DMSP spacecraft, but not over POES spacecraft, which are still operated by NESDIS under the authority of the NOAA/NESDIS Office of Satellite Operations. Eventually, the NESDIS polar program office will be brought under the same roof as DMSP in the IPO, so that all polar weather satellites are truly a converged system. Also, five more DMSP and four more POES spacecraft, already contracted and built, will be launched to maintain the existing constellation until NPOESS spacecraft are available around 2008. Figure A.1 illustrates the current and future NPOESS constellation.

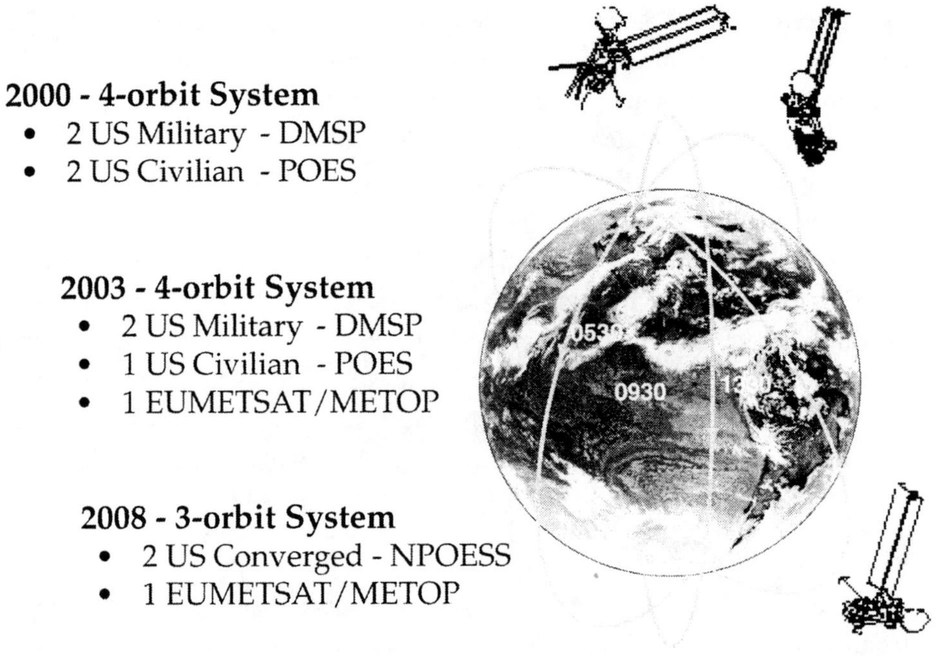

2000 - 4-orbit System
- 2 US Military - DMSP
- 2 US Civilian - POES

2003 - 4-orbit System
- 2 US Military - DMSP
- 1 US Civilian - POES
- 1 EUMETSAT/METOP

2008 - 3-orbit System
- 2 US Converged - NPOESS
- 1 EUMETSAT/METOP

0530
0930

Source Material: 21 Oct 99 NPOESS briefing to AIAA by John Cunningham, SPD

Figure A.1 Present and Future NPOESS

As Figure A.2 below indicates, the organizational structures need further development to streamline operations and fully realize convergence. While the Office of Satellite Operations (OSO) handles operations for GOES, POES, and DMSP, program management of POES and DMSP is split between two offices. As the dotted line figure implies, POES program management will eventually move from the Office of Systems Development (OSD) to the NPOESS IPO

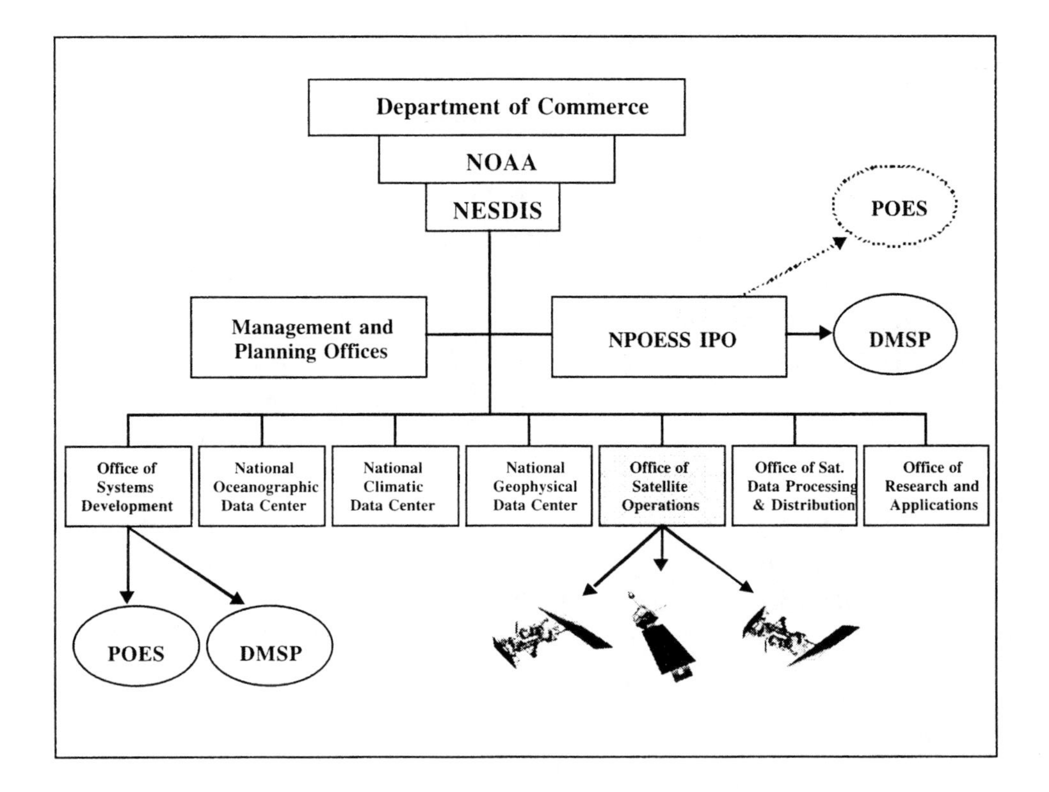

Source Material: NESDIS website http://www.nesdis.noaa.gov/

Figure A.2 NPOESS Organization Within the Department of Commerce

NPOESS Case Study Approach

This case study examines the early history of the NPOESS IPO and primarily investigates how the Departments of Defense and Commerce and their respective DMSP and POES program offices implemented convergence and formed an IPO. RAND expected to find many valuable lessons learned from the experience, which can be applied to other programs involved in creating joint program offices. Although full convergence is not yet complete, the NPOESS IPO has managed a significant change in how present operations are carried out and how the entire U.S. meteorological program will look in the future.

While the NPOESS IPO has had a relatively brief history, valuable lessons in organization, management and operations are still evident and offer educational insights. This case study focuses on the early years of the IPO, from roughly 1994 to 1998, where program managers had to develop plans, agreements, and requirements in accordance with policy and federal regulations.

RAND took a two-pronged approach to the NPOESS IPO case study. First, the key program documents were examined in roughly chronological order to provide a framework for how the IPO organized. The documents included the Implementation Plan, Memorandum of Agreement, and Requirements Documents. There are a number of critically important documents that any joint program must successfully complete not only properly, but also early. Second, examination of the program documents was complemented by interviews with NPOESS IPO personnel to gain their observations of the program. Expectations were that the insights of program personnel might reveal many lessons that cannot be gleaned solely from analysis of the documents.

NPOESS IPO Documents and Process

Implementation Plan for NPOESS – May 2, 1994

In 1993, prior to the presidential announcement of NPOESS, a Triagency Convergence Study Group[86] was commissioned by OSTP to come up with an implementation plan to determine how convergence could be achieved. The Implementation Plan described how the program would operate, how assets would be merged, and how authorities and responsibilities would be delegated. The Plan was a "big picture" overview of NPOESS in 1994 and was written before the IPO existed. It did not contain extensive details on many points, which are better explained in later documents that this plan calls for. The importance of the Plan lies in showing that extensive programmatic pre-planning is very necessary, rather than leaving such details until later. Other key points from this Implementation Plan include:

IPO Structure: The Triagency Group specifically chose an IPO approach among three others, including Single Agency, Dual/Distributed, and a Government Corporation. The IPO approach "maximizes use of the technical expertise provided by the participating agencies under a single System Program Director who will provide a coordinated programmatic focus."[87] After studying lessons learned from other interagency programs, including Landsat 7, the study group determined that the following conditions were essential for a successful IPO:

- Each agency must be institutionally committed to the success of the program.

---

[86]The Study Group consisted of members from DoD, NOAA, and NASA. The members worked for about a year to produce the Implementation Plan.

[87]*Implementation Plan for a Converged Polar-Orbiting Environmental Satellite System*, White House Office of Science and Technology Policy, May 2, 1994, p. 3.

- There must be a single program manager with the authority and responsibility to manage all converged system activities across agency boundaries.
- Collocating representatives from all stakeholder agencies into an integrated program office under the direction of a single converged system manager greatly increases the chances for success.
- Increased joint involvement of agency participants in day-to-day activities and problem solving within a single organizational structure increases coordination and chances for success.
- The requirements baseline must be defined before the acquisition begins; that baseline must be configuration-controlled and any changes to the baseline must be made only after careful consideration by senior leaders in the agencies involved.
- All funds for the program should be managed and defended by the Integrated Program Office.[88]

Architecture: The Plan clearly defines the most important aspects of the combined systems architecture, including policies regarding space and ground assets and data distribution. As the report states, "The system will be open in character. The system also has an important requirement to selectively deny real-time critical environmental data to an adversary during crisis or war, yet ensure the use of such data by U.S. and Allied military forces."[89] It is clear that the interests of both DoD and NOAA had to be clearly enunciated before either would commit to relinquishing control of their respective programs.

Even though DMSP and POES were very similar, the actual satellites carried some distinctively different hardware and served different mission requirements. The Triagency Group had to find the best strategy to bring both assets together under one operational roof. They commissioned a series of studies to determine how both types of satellites could satisfy all mission requirements. The Plan concluded that the current satellites then under construction for DMSP and POES could not be significantly redesigned without suffering intolerable cost overruns. The new NPOESS would simply fly out the current two programs until they could be replaced with a new generation of assets specifically built for the integrated program.

---

[88] *Implementation Plan*, pg. 2

[89] *Implementation Plan*, pg. iii

Budgeting: The Implementation Plan clearly indicates that the best solution for budget policy would be single agency funding, since it removes all potential problems inherent with multiple financial sources. However, all recognized that this was not feasible in 1994 and it has still not occurred in 2000. Nor will this situation likely change in the foreseeable future, according to IPO leaders. Even though single-agency funding was not possible, the Plan does specify that the IPO will have primary responsibility for the development and justification of budget planning estimates, thus centralizing the coordination process. Also, since convergence was expected to realize impressive cost savings, the authors stress this point in the Plan and promised detailed reports to both OMB and OSTP to show how savings come about over time.

Civilian Leadership: From the outset, a converged program was to have a distinctively civilian image, most reflected in how leadership positions would be allocated among the agencies. The Plan states,

> The System Program Director will be an employee of NOAA reflecting the inherently civil nature of the program and to provide the formal programmatic interface with international partners and users. Inherent in this overall management function is the responsibility to ensure the program execution of the U.S. policies (within the new constraints of the single converged U.S. meteorological polar orbiting satellite system) of open civil distribution of meteorological data and products for the global community.[90]

The military was expected to retain significant influence over the converged program in management and operations, but the Plan made it clear that the SPD would lead all interactions with Congress and foreign nations.

Transition Activities: The Triagency Group charted a specific set of actions and timelines to initiate NPOESS convergence. In concert with OSTP, they drafted the Executive Order used in the Presidential Directive. To establish the IPO, they drafted the first Memorandum for each agency to sign, while also initiating development for the first Integrated Operational Requirements Document (IORD).

Memorandum of Agreement (MOA) Between DoC, DoD, and NASA – 26 May 1995

Just over a year after the NPOESS presidential directive, the three agencies signed an MOA, stating that this document "constitutes the formal agreement, including roles and responsibilities, between DoC, DoD, and NASA..."[91] Many

---

[90] *Implementation Plan*, p. 8.

[91] *Memorandum of Agreement Between the DoC, DoD, and NASA for the NPOESS*, p. 1.

parts of the MOA read like the Implementation Plan described above, yet the MOA contains significantly more details that are binding upon the signing agencies. The character and management structure of NPOESS are laid out in explicit detail, pursuant to specific regulations of authority, making each agency recognize that funds, responsibilities, and personnel are now committed to action. The MOA can properly be called the cornerstone document of NPOESS, which covers:

1. Organization
2. Responsibilities
3. Requirements
4. Management and Process
5. Effective Date and Amendment/Termination policies

Each are described in greater detail below.

1. Organization: The MOA organization plan is very similar to recommendations made in the May 1994 Implementation Plan. As shown in Figure A.3 below, the NPOESS IPO receives direction from four committees, each staffed with representatives from all three agencies. Organizational responsibilities detailed in the MOA are as follows: DoC is primary for satellite operations and running the IPO, DoD heads acquisition, and NASA coordinates technology issues. DoD and DoC are the primary players in the cooperative effort, while NASA plays more of a support role. The four committees' key members and functions are:

- Executive Committee (EXCOM):
    - DoC: Under Secretary of Commerce for Oceans and Atmosphere
    - DoD: Under Secretary of Defense for Acquisition and Technology
    - NASA: Deputy Administrator
    - Function: Provide policy guidance, ensure sustained agency support (including funding), endorse the NPOESS requirements baseline, review the annual business plan, and approve:
        - annual budgets
        - NPOESS staffing plan
        - acquisition program and major changes
        - modifications or waivers to existing NPOESS policies

    - Convergence Master Plan

- Joint Agency Requirements Council (JARC):
    - DoC: Deputy Under Secretary of Commerce for Oceans and Atmosphere
    - DoD: Vice Chairman of the Joint Chiefs of Staff

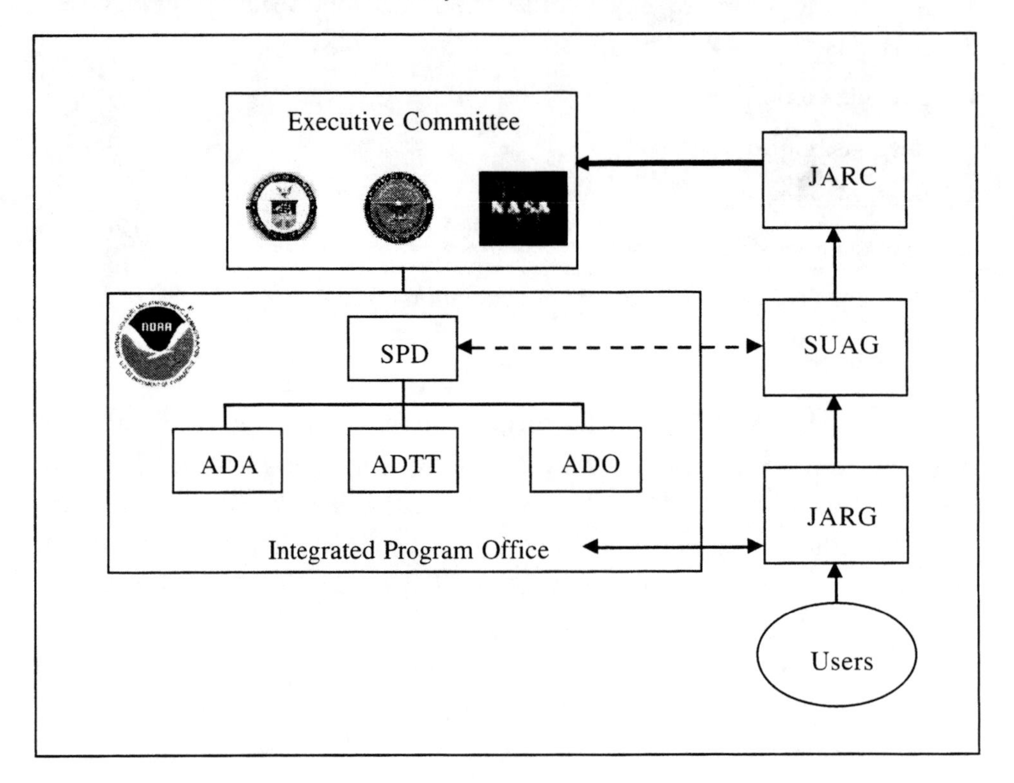

EXCOM = Executive Committee

JARC = Joint Agency Requirements Council

SUAG = Senior Users Advisory Group

JARG = Joint Agency Requirements Group

Source Material: NPOESS MOA

Figure A.3 NPOESS Organization

    - NASA: Associate Administrator for Earth Science Enterprise
    - Function: Approve the final NPOESS IORD and resolve documented interagency requirements disputes not solved at lower levels

- Senior Users Advisory Group (SUAG):
    - Members include a limited number of NPOESS user agencies within DoC, DoD, and NASA. The SUAG is independent of the IPO and

directly represents users in the U.S. government at large. SUAG chairmanship rotates biannually between DoC and DoD. No single agency will simultaneously chair the JARC and SUAG.

- Function: Represent U.S. Government NPOESS user concerns and advise the IPO System Program Director on user community needs and how program decisions relate to the satisfaction of the IORD.

- Joint Agency Requirements Group (JARG):

  - Members come from several offices within DoC, DoD, and NASA, including Air Force Space Command, National Weather Service, Goddard Space Flight Center, National Marine Fisheries Service, and many more. The JARG chairmanship rotates biannually between DoC and DoD.

  - Function: Develop the NPOESS IORD and administer the approval process. Develop a requirements master plan for JARC approval, using DOD policies and procedures as a basis.

It was very important to planners that NPOESS have a single voice and as much autonomy as possible in regard to program management. As stated in the MOA,

> The IPO, under the direction and management of the SPD, will be the single functional entity responsible for the planning, budgeting, development, acquisition, launch, operation, and management of the NPOESS. The SPD is ultimately responsible to the triagency EXCOM for NPOESS. The SPD has decision authority for NPOESS matters, subject to the statutory authorities of the designated agencies, and reports to the NOAA Administrator . . .[92]

The System Program Director (SPD) reports through an operational chain within DoC, including the Associate Administrator of NESDIS, then to NOAA, and finally to DoC, a hierarchy similar to the POES program. The SPD also coordinates matters that affect DoD through the Assistant Secretary of the Air Force for Space.

In the MOA, the IPO was designed to have three functional line offices and a SPD support staff, as shown in Figure A.3. The developers of NPOESS specifically designated responsibilities and appointment authority of the SPD and the line offices to fully represent the interests of each primary agency.

- System Program Director (SPD):

  - Direct NPOESS and be responsible for financial, programmatic, technical and operational performance of the NPOESS.

---

[92]MOA, p. 3.

- Associate Director for Acquisition (ADA):
    - Responsible to the SPD for developing, acquiring, and fielding the NPOESS components and for launch and early on-orbit checkout.
- Associate Director for Operations (ADO):
    - Responsible to the SPD for NPOESS operations, which includes command, control, and health of spacecraft, acquiring telemetry, ensuring communications, anomaly support, mission planning, and user interface.
- Associate Director for Technology Transition (ADTT):
    - Responsible to the SPD for promoting transition of new technologies that could cost effectively enhance the capability of NPOESS to meet operational requirements.

2. Responsibilities: The MOA designates lead agency responsibility for a number of tasks, to ensure there is no confusion over authority and to support the primary philosophy of the NPOESS effort. DOC handles operations and administration, DOD leads acquisition, and NASA coordinates technology issues. Even if one agency has the lead, it still receives guidance and manning from the other two.

- DoC/NOAA
    - Nominate SPD, who is approved by EXCOM
    - Satellite and ground operations
    - Lead user-interface agency
    - Nominate ADO, deputy ADA, and provide majority of SPD staff
- DoD
    - Lead NPOESS acquisition
    - Nominate deputy SPD, ADA, deputy ADO, deputy ADTT, and provide majority of acquisition personnel
- NASA
    - Lead agency to develop and insert new cost-effective and enabling technology, in conjunction with Earth Science Enterprise assets
    - Nominate ADTT and provide necessary personnel to support each NPOESS directorate

3. Requirements: The NPOESS IPO used most aspects of traditional acquisition programs and therefore, DoD acquisition requirements are found in the IORD. The MOA states,

> An IORD will be the sole operational requirements source from which triagency cost and technology assessment, specification development, and related acquisition activities will be conducted. The requirements process will be independent of the IPO and is designed to ensure each agency's requirements are accountable and traceable to each agency.[93]

NPOESS IPO wanted to take advantage of DoD's extensive experience with acquisition of highly technical, expensive programs, based upon processes outlined in the 5000 series instructions. It is important to repeat that each agency is fully represented in the process, even though DoD led the effort. The JARC and JARG were especially created to fill this objective. Figure A.4 illustrates how NPOESS requirements were organized, consistent with DoD acquisition processes.

The MOA requirements section also includes preliminary budget outlays expected from the lead agencies DoC and DoD (see Table A.1 below). All IPO leaders strongly emphasized that it is vitally important that all parties recognize their financial obligations as early in the program as possible. While the actual numbers have proven quite different from the original budgets, it is important to note that DoD and DoC agreed to share costs equally.

4. NPOESS Management and Process: The MOA describes a very ambitious plan to structure the management of the NPOESS IPO. The SPD and staff were identified as responsible for an annual business plan as well as long-range staffing plans, which addressed the key issues in day-to-day business and how the program would meet expected milestones. These two plans are fundamental to a traditional program and have been utilized as expected.

An early effort described in the MOA was the Convergence Master Plan (CMP), a very detailed set of plans that the NPOESS organizers expected to meet a wide array of managerial needs during convergence. The CMP was designed to contain the following:

---

[93]MOA, p. 15.

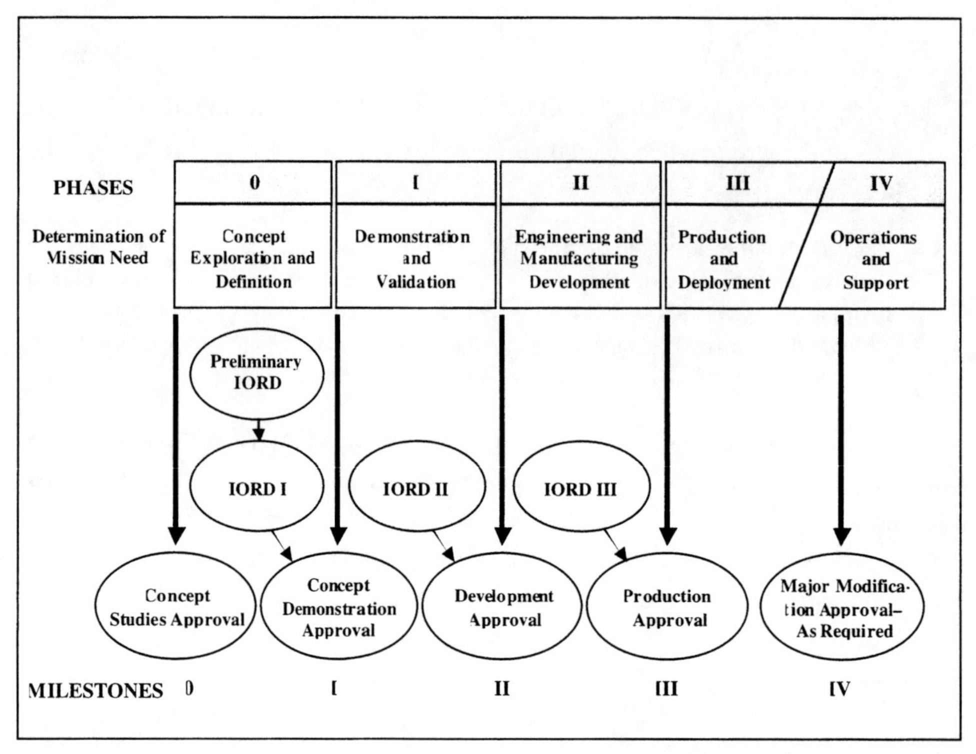

Source Material: NPOESS MOA

Figure A.4 NPOESS MOA Program Acquisition Plan

	FY96	FY97	FY98	FY99	FY00	FY01	TOTAL
BUDGET ($mil)	78.0	120.0	187.0	340.2	372.7	328.1	1426.0
DoC	54.0	78.2	131.4	146.5	162.5	140.4	713.0
DoD	24.0	41.8	55.6	193.7	210.2	187.7	713.0

Source Material: NPOESS MOA

Table A.1 NPOESS MOA Preliminary Budget

- Acquisition Management Plan
- Technology Transition Management Plan
- Operations Plan
- Funding Management Plan

- Integrated Organizational Management Plan

The MOA included lengthy instructions on the contents and timelines of each plan within the CMP. The first draft of a CMP would have been due six months after the SPD was appointed, however it became clear early on to IPO personnel that the CMP was too ambitious and in many ways, completely unnecessary, since many functions of the document were covered elsewhere. Some parts of the CMP gained attention, but a completed CMP does not exist. The IPO staff does not expect that it will be necessary to complete the CMP.

5. Effective Date and Amendment/Termination Policies: The authors of the MOA felt it important for the cooperating parties to understand and document dates of implementation, methods of making changes, and policies regarding termination of the cooperation in case events change. In the NPOESS MOA, these terms were agreed upon and signed by the Secretary of Commerce, Secretary of Defense, and the NASA Administrator.

Integrated Operational Requirements Document (IORD) I – 28 March 1996

NPOESS is designated a DoD Acquisition Category (ACAT) Level I program subject to DoD Directives 5000.1 and 5000.2, as well as normal scrutiny to the requirements of OMB. As stated in the May 95 MOA, the IPO was required to produce an IORD to fully describe the requirements of the NPOESS program. From Figure A.4 above, we can see that IORD I was only one in a series of requirements documents. It is the primary "meeting place" where each agency participating in the IPO ensures their requirements are addressed. As the current NPOESS SPD noted, "Converged requirements provide foundation for the converged program."[94] The NPOESS IPO had extensive negotiations between members and within the committees to finally complete and approve the IORD. IPO personnel agree that a program should not proceed unless there is significant agreement within the IORD or similarly utilized document. A failure to agree upon fundamental requirements early will become more evident later when allocating budgets and meeting mission requirements.

An IORD for any technical program is a very lengthy, complicated document. This study will not provide any significant details on the NPOESS IORD, except to outline the most fundamental elements.

NPOESS IORD:

[94] 21 Oct 99 NPOESS briefing to AIAA-John Cunningham, SPD.

- Function – The sole operational requirements source to conduct cost and technology assessment, specification development, and related acquisition activities.
- Defines:
    - Operational capabilities and mission needs
    - Threat and threat environment
    - Shortcomings of existing systems
    - Capabilities required in space, ground, and user segments
    - Integrated logistics support
    - Infrastructure support and interoperability
    - Force structure
    - Schedule considerations

Other NPOESS Documents

After the Presidential Decision NTSC-2, the Implementation Plan, the 95 MOA, and the IORD, there are several other documents that have led to successful NPOESS convergence. Many are a result of the transfer of authority between programs, agreements on budgets and operations, or required administrative provisions to facilitate IPO authority. The following list of documents, though far from complete, should give an indication of efforts required for satisfying the varied needs of multiple parties in a convergence.

- C3 Concepts of Operations (CONOPs) for NPOESS, 18 Oct 95
- Implementation Plan for the Transfer of DMSP Operations to the NPOESS IPO, 25 June 96
- MOA for DMSP Satellite Deployment and Support between AFSPC 50th Space Wing and the DMSP System Program Office, 25 Feb 94
- MOA for DMSP Operations Funding between HQ USAF/AQS, HQ USAF/XOF, HQ AFSPC/DO, and the NPOESS IPO, 21 Nov 96
- DMSP Operations Transfer Criteria Document, between NPOESS IPO and AF Space Command, 7 Mar 97
- Employment Plan for DMSP Operations at the Suitland Satellite Operations Control Center (SOCC) at Suitland, MD and the Environmental Satellite Operations Center (ESOC) at Falcon AFB, CO, 18 Nov 97

- MOA on the Deployment, Employment, and Sustainment of the DMSP between the NPOESS IPO, AF Weather Agency, US Space Command, and AFMC, SMC/CI, and DMSP System Program Office, 16 Nov 99

Interviews with NPOESS IPO Personnel

RAND was fortunate to have discussions with a number of NPOESS IPO personnel in order to gain additional insights into IPO formation. An extensive lead-in session involving approximately 15 IPO personnel, including the SPD, ADO, and other key staff representing NOAA, DoD, and NASA was conducted in December of 1999. Also, RAND carried out a number of follow-up interviews to further clarify issues raised from the lead-in session.

The IPO personnel took advantage of the occasion to discuss long-standing issues that have accumulated in their first six years of operation. While the NPOESS IPO can generally be called a success, they have endured their share of problems in all aspects of operations and management of a complex system within the federal bureaucracy. The session revealed insights that may not be found in any "JPO manual." These have been distilled into a number of broad categories of "lessons learned" that could be applied to similar joint programs within the U.S. government.

- Initial Organization and Structure
- Implementation Plan
- MOA
- Early Convergence
- Requirements
- IORD
- Stakeholder Support
- Management and Responsibilities
- Budgeting and Authority
- Open Access vs. Security
- Staffing

Initial Organization and Structure

Organizers had tried to form a joint military-civilian program many times, but it only occurred in the form of NPOESS on the ninth try in the early 1990s. Previous failures resulted from an inability to reconcile requirements and meet mission needs of the DoD and civilian meteorological community. However, with the end of the Cold War and an impetus toward fiscal austerity, NPOESS finally arrived.

A key component of early success was lining up top level support in all areas that had influence upon the decision to integrate programs. Having this support served to deflect interference at lower levels. In the Executive Branch, OSTP and the President fully endorsed NPOESS, resulting in the Presidential Directive. Military leaders at the four-star level and equivalent civilian managers at DoC and NASA were brought on board to ensure NPOESS had significant backing, even though many people at lower levels, especially in DoD, still had concerns about whether a merged program would continue to meet their needs.

Initially, there existed only a thin veneer of trust between DoD and DoC, stemming from fundamental differences in mission, culture and management approach. DoC/NOAA and NASA are mostly concerned with open access, scientific study, and long-term, focussed research, while DoD emphasized warfighter support and continuous, near real-time operations. Of key interest to military leaders were asset and data distribution controls. Could the NPOESS assets meet national security needs effectively and efficiently in the same way as DMSP?

Also, DoD was wary of having NPOESS IPO direction under a civilian SPD, who might not devote sufficient attention to national security concerns. As reflected in the Implementation Plan, national leaders wanted NPOESS to have a distinctly civilian flavor, to allay fears of a military takeover of POES. This concern and others were addressed early in signed documents and accepted by top DoD leaders, reinforced again by the Presidential Directive. As it turns out, SPDs and many other key NPOESS leaders, while civilian members of NOAA or NASA, have some measure of military experience and have emphasized meeting the interests of all three agencies. All IPO personnel agree that it is critically important that the SPD be acceptable to all agencies, and not be a political appointee with limited or narrow experience.

Implementation Plan

Convergence had failed in the past due to irreconcilable differences in many areas. It was much easier to justify separation in the 1980s, but after the Cold War, pressures caused by reduced budgets and less clearly defined threats provided incentives for convergence, despite concerns on aspects of how convergence would occur. The Implementation Plan thus became a document created by people from each agency, willing to work out differences and guarantee future weather missions.

Memorandum of Agreement

The Implementation Plan was an important first step toward realizing NPOESS, but it represented no actual commitment of agency resources. Both DMSP and POES continued to operate separately in early 1995, while convergence planners continued to develop NPOESS plans and documents. The MOA is seen by many as the pivotal agreement whereby the idea of an NPOESS gained sufficient momentum to make the program inevitable. This is where top leaders in each agency agreed to a concrete plan that committed budget, personnel, and hard assets in a new direction.

It is important to note that the original MOA from 1995 has not been amended in the last six years, a testament to how well it was constructed. Follow-on MOAs have primarily dealt with very specific issues related to operational support, temporary budget measures, and refining some responsibilities.

The NPOESS MOA was significant for a number of reasons, besides being a cornerstone of the program.

- As described above, NOAA and DoD did not fully trust one another. The MOA was written to specifically address all critical questions where lack of trust arose, including statements on mission and data distribution, authority, and responsibilities.
- Primary budgets were laid out, so that neither party could later question basic fiscal responsibilities.
- Key leadership positions and committees were described in detail, ensuring each agency had representation to address current and future concerns. Even at the signing of the MOA, DoD and some others still had real concerns over NPOESS, but within the document, they had the means to continually address those concerns. The MOA was written to be very explicit in most ways, as described above, while some areas were left to interpretation.

NPOESS had to follow basic DoD acquisition rules, but those rules had enough flexibility to utilize "best practices" in NASA and NOAA to meet their needs. Also, while convergence was mandated, the exact manner and timelines of convergence were not, allowing the agencies to combine their assets within realistic scenarios. Many would insist that POES should already be under the NPOESS roof, even before DMSP, but it did not occur for a number of reasons. Yet, IPO leaders are confident that it will happen, even if on a timeline that does not suit everyone.

- Since NOAA was designated lead agency, operations and international interface went to this agency. DoD, which has significant acquisition experience, received lead in this area, which also helped calm some fears in DoD that future systems would not meet DoD missions.

- The MOA describes a Convergence Master Plan in detail, a legacy of DoD CONOPs requirements regulations. The IPO quickly learned that the CMP was redundant and was dropped very early. Nobody has missed it nor called for anything similar since.

Early Convergence

The NPOESS IPO was officially created on 1 October 1994 and DMSP operational integration occurred on 26 May 1998. Convergence is a continuing process in 2000, while full system integration is not expected until at least 2008. From the first years of IPO operations, NPOESS IPO personnel noted several important factors.

- Each agency had representatives within the IPO. All of them approached the IPO mission with optimism and dedication, despite some of the nagging issues noted below, much because the program had already received affirmation at high levels. Each agency performed their duties as required, while no significant levels of dissension existed that could have upset early IPO performance.
- IPO organizers purposely collocated all personnel in a centralized office, in order to develop team cohesion through daily interaction and cooperation. Also, as much as possible, activities were carried out by integrated teams, so that isolated groups would not form.
- The process of integration was drawn out over several years, so that large gaps of time existed between key events. IPO managers noted that they were able to maintain momentum by creating a "bandwagon effect" early, which helped to keep the process moving.
- Involved members tried to instill a deep sense of ownership among all NPOESS personnel, so that IPO success became a group goal. The backup operations facility was located at Schriever AFB and NPOESS DMSP continued using the AFSCN assets, which immediately gave the military a sense of joint ownership, opposing fears of a civilian takeover.
- It is very helpful to have skilled, committed leaders, champions if possible, with direct influence in the Executive Administration and Congress. NPOESS did fairly well in this area, but had budgetary problems as noted below.

Requirements

Systems and operational requirements are the glue that ensure the integrated program stays together. It is essential to agree on all important requirements early, long before acquisition, in writing, and with the flexibility for new requirements later.

Requirements were the key stumbling block to DMSP-POES integration in all the earlier attempts. The NPOESS agencies approached the concept quite differently; DoD was very formal and desired significant detail so they could ensure technical data met mission needs, thus avoiding "requirements creep," and could mitigate the effects of high personnel turnover rates. NOAA and NASA, however, preferred to rely more on general goals or statements, leaving excessive detail to support documents created at a later date. The civilian agencies normally do not have high personnel turnover, so they are confident that their corporate knowledge is sufficient. While DoD organizes requirements in specific regulations and validates them through the JROC, the National Academy of Sciences has no equivalent. NASA and NOAA utilize an evolved process via technology push.

A necessary obligation of the requirements process in federal programs is submitting plans to budgetary analysts. OMB, DoD, and DoC comptrollers must review the NPOESS program to make sure taxpayers get their money's worth. In the process, the IPO must continually educate budget officers and defend specific requirements, as well as the overall program. Turnover rate in these outside agencies only makes the problem more pressing.

Budget officers are fond of cost/benefit analyses, especially within DoD, where they like to understand the mission impact of a system or even component systems. The more one can initially quantify requirements into explainable results, the more one can provide concrete justification and defend details and costs of each element in the program. This becomes a more difficult problem when dealing with civilian environmental requirements, where needs are not as easily quantified as weapons systems. IPO managers emphasized the need to quantify requirements as early as possible, so they could be defended to auditors with a maximum level of justification for the system in question.

IORD

The requirements process became a longstanding and heated issue in the early days of NPOESS, but never a showstopper, because all parties managed to work out their differences. Scientists could not understand the need for so much detail in the IORD, and there existed much debate over terminology and technical details. For every measurement or specific metric, the group had to find acceptable thresholds, wherein each agency's technical specifications and tolerances would fit. Since convergence was already mandated, these disputes simply had to be worked out, so it became a matter of keeping everyone negotiating until an IORD was produced. DoD led the acquisition effort, so the requirements product would exist within the IORD format, but lesser details had to be worked out after much effort.

Stakeholder Support

Few programs that draw significant federal funds can thrive simply upon the momentum of the initial idea. Program advocates at the highest possible levels must continually sell and defend the program in several forums, including Congress, the Executive, key agency meetings, and open conferences.

Following the Presidential Directive creating NPOESS, the program had enough momentum to carry through the first few years of IPO formation, but that momentum could not be maintained. Every year, the IPO faces significant budget cuts and must continually educate auditors on the details and necessity of program funding. Not only the SPD, but also leaders in NOAA, NASA and the Air Force have championed NPOESS over the years. The IPO staff suggested several obvious, but important strategies to ensure support of a joint program.

- Target key members of congressional committees that influence the program. For NPOESS, this includes leaders in Authorizations and Appropriations of the House and Senate, both on the Defense and Commerce Committees.
- Maintain support upward in the chain of each agency. Directors of NOAA, DoC, and NASA, and senior executive and flag officers in DoD must provide unwavering support.
- The IPO is expected to brief OSTP annually on NPOESS status. The IPO tries to take advantage of all opportunities to maintain Executive support.
- OMB comptrollers should receive the best possible information on the details of the program. Educating OMB and other comptrollers in DoC and DoD has remained a full time job for some in NPOESS, especially the SPD.

Management and Responsibilities

As previously noted, there have been no significant changes to the original MOA and structure of the NPOESS program. The IPO and the support committees (EXCOM, JARC, SUAG, and JARG) have essentially carried out their prescribed tasks, but some comments are worth noting from their first six years of experience. A common problem of each committee has been a high rate of turnover, so that, on average, members often do not have the necessary knowledge to be most effective. The problem is most common with military members, but also to a lesser degree with civilians. IPO managers stress that they must be always ready to continually educate those who have influence upon their operations.

EXCOM: This committee was expected to operate in the fashion of a board of directors, providing senior level program guidance on big issues such as budget, IPO leadership and strategy. On average, the EXCOM has met about twice a year. Members have on occasion been called to champion NPOESS in appropriations issues in Congress and at high levels of each agency.

JARC: This committee has had the least utility of the four. Its primary purpose was to approve key requirements such as the IORD and resolve major disputes between agencies. Many disputes existed among agencies, but they never rose to levels requiring JARC attention. The JARC has only met once.

JARG: The JARG has been a consistently active committee, which met up to twice a week in the early years of the IPO, tapering off to monthly or even quarterly meetings when requirements issues were not as pressing. Members of the JARG consisted mostly of military officers at the O-3 and O-4 level and civilians of comparable position who determine and adjudicate detailed requirements. While the interests of the agencies were represented well enough, some IPO personnel noted that all key user interests might not have been. It is important that requirements meetings include the "customer," since their requirements eventually determine success of the system.

SUAG: Whereas the JARG could be called the middle management requirements group, the SUAG, as the name implies, contains senior level leaders who advocate requirements to meet user needs. Based upon the phase of the program, the SUAG has met as little as once a year up to once each quarter. IPO managers mostly agree that this committee has served a valuable purpose in representing users and providing input to the requirements process.

Budgeting and Authority

Maintaining financial resources for the NPOESS converged system has been the primary, continuous problem for IPO managers. They estimate that budgetary matters can often fill roughly 80% of their work hours, which includes educating agency and congressional staffs about budget and program requirements.

The NPOESS program is expected to cost nearly two billion dollars up to 2006 for sensors, satellites, and ground system development and construction, as well as for IPO operations. Like any other large program spread out over several years, NPOESS is constantly subject to program cuts within discretionary funding debates. The fiscal constraints that helped inaugurate the program have also caused delays and cutbacks. The IPO must weigh all their alternatives to meet various scenarios as lawmakers annually threaten to reduce NPOESS funding. Some sensor suites may be reduced in scope or cancelled, or more likely, deployment of the entire spacecraft is delayed.

An important problem that the IPO could not avoid was split budgeting between DoD and DoC/NOAA. Managing one budget is difficult, but NPOESS must get their monies and argue their priorities within two separate bureaucracies. They must navigate two different budget processes and deal with different representatives at OMB. Also, DoD and DoC/NOAA have different budget cycles, so that information is due to comptrollers at different times of the fiscal year. Since the program still relies on 50/50 funding, it is important and difficult to ensure that consistent numbers are used. Annually, the IPO must clear the following budgetary hurdles:

- NPOESS has traditionally not had problems with its budgets at the NESDIS or NOAA level. However, upon reaching the DoC comptrollers, NPOESS budgets receive much more scrutiny, as all programs compete to fit under departmental budgetary caps.
- In DoD, NPOESS funding falls under the Air Force. The budget cycle begins a few months before DoC's in February, a process that is more bureaucratically demanding, especially since NPOESS must compete with a wider range of DoD programs. It has been an annual challenge to submit coordinated budgets to DoD and DoC, so those comptrollers can examine consistent funding numbers. Normally, the IPO has had few significant problems with DoD funding, except in maintaining funding at the 50/50 pace when DoC considers cutbacks.
- Within OMB, NPOESS must coordinate with three separate auditors for DoD, DoC, and NASA. These comptrollers normally coordinate their activities, which means that NPOESS must always be consistent in providing information to outside agencies.
- In Congress, NPOESS funding must face multiple House and Senate Authorizations and Appropriations Committees. For DoC, the key Appropriations Sub-Committee is for Commerce, Justice, State and Judiciary, while for DoD, it is the Defense Sub-Committee. Other congressional committees, such as the House Subcommittee on Science, Technology, and Space, also look at NPOESS, but these are not as important to the fiscal survival of the program.

The IPO must therefore lobby their cause among various legislative and departmental bodies that have distinctive political, mission, programmatic and budgetary priorities. Single source budgeting would go a long way to alleviating many of these problems, but dual-source funding is something the IPO must live with for the foreseeable future.

Within DoC, NPOESS is a key program with high visibility and a budget that will grow to represent around 25% of the NOAA budget. Any changes to the program send shockwaves through the agency and easily get the attention of top leaders. Problems with a relatively large program can be worked out quickly within NOAA and DoC. In DoD, the situation is reversed, since NPOESS is only a small part of even the space budget. The program is not a top priority among most Pentagon leaders, so problems take much longer to work through the larger DoD bureaucracy.

Since single-source funding may not be a reality for NPOESS, the IPO must continue to operate as usual. Like the requirements process, IPO managers stress that funding rules and procurement timelines must be set up and agreed upon early.

A common response from the IPO staff is that they were sometimes affected by decision bodies where they have little or no influence, let alone representation. NOAA and the IPO SPD have worked for more influence with governing space bodies, such as the Partnership Council,[95] which can have an impact upon NPOESS. A valuable lesson from this experience is to assure the program has a seat in any forum or council that has real impacts upon the program, especially in budgetary matters.

Related to this is that NPOESS cannot work out high level problems within any centralized adjudication body, such as the National Space Council.[96] Program leaders need a forum to take their issues, so that disputes can be worked out in one place. Much of this arises again from the fact that NPOESS is split between NOAA and DOD, while space matters often transcend both.

---

[95]The Partnership Council is a formal relationship between Air Force Space Command, NASA, and the NRO. It was established in February 1997 with a goal towards initiating cooperative efforts to save money, reduce risks, and integrate planning efforts in areas of mutual interest. Issues for possible collaboration include launch range modernization, launch infrastructure and support activities, preparation for space activities such as the Leonid Meteor Shower, and space technology cooperation. See NRO Press Release, "NRO Joins NASA, Space Command in Partnership Agreement," 21 December 1998.

[96]The National Space Council was reestablished during the Bush Administration as an interagency mechanism to address cross-cutting space-related issues and programs. The Clinton Administration chose not to use the formal mechanism of the National Space Council but to address space issues in OSTP, the National Security Council (NSC), the National Science and Technology Council (NSTC), and elsewhere.

Open Access vs. Security

A suspected large barrier to integrating DMSP and POES was whether a combined system could satisfy both agencies' data access policies. DoD has top secret clients who require significant support within a tight security framework. NOAA, on the other hand, rarely limits distribution of data, especially among friendly nations.

During convergence, this problem did not end up as difficult as expected. DoD has been able to continue high-quality service to national security customers while also maintaining security. Also, selective access of data is still possible, though not normally required. NOAA, with POES, did have experience with selective access during the Falklands War, while during the Gulf War, the problem was not denial, but rather not having enough coverage.

DoD has learned to operate with greater flexibility, while NOAA has adjusted to limited selective access. Undersecretaries of Defense and Commerce meet on a continuing basis to resolve any issues in this area, but to date, open access and security have coexisted within NPOESS without any serious problems.

Staffing

Over the first six years, staffing for the NPOESS IPO has grown, and presently stands as shown below. The IPO has been able to complete their most important tasks while "undermanned" for a number of reasons, including hard work by many personnel. Also, in the next decade, as NPOESS begins to build and deploy new assets, more personnel will be required.

It should be noted that DoD supplies a proportionally larger share of the IPO personnel. Many of the civilians, like the SPD, are retired or reserve military, which has been a major factor in giving DoD the confidence that NPOESS will continue to meet critical national security weather missions.

Tables A.2 and A.3 below show authorized and assigned personnel within the NPOESS IPO. The acquisition team, and the IPO in general, are small compared to similar programs in DoD that deal with such expensive hardware and missions. IPO managers generally feel that smaller is better, since the teams can be more flexible and timely in carrying out tasks. Also, roughly 50 civilian contractors interact with IPO on a daily basis. Most of them work on issues related to new satellite design, especially the sensor suites, and mission requirements.

The NPOESS IPO has been a dynamic organization that has had to become very flexible over the last six years. Overall, they have accomplished their essential

Agencies	Filled/Assigned	Authorized
NOAA	16	22
NASA	6	6
Air Force	25	36
Navy	4	6
Army	1	1
Total	**52**	**71**

Source Material: NPOESS IPO Organization Chart, January 2000

Table A.2 Assigned and Authorized Personnel by Agency

Program Director (SPD)	(29)
— SPD office	7
— User Liaison	4
— External Affairs	2
— Program Control	6
— Systems Engineering	5
— Contracts	5
Acquisition (ADA)	22
Technology (ADTT)	6
Operations (ADO)	14
Total Authorized to IPO	**71**

Source Material: NPOESS IPO Organization Chart, January 2000

Table A.3: Authorized Personnel in IPO by Section

goals, but also have had several small problems that have hindered greater success.

- The position of SPD has understandably become very important to the IPO. Even so, the program has had to operate for long periods without an officially appointed SPD, due to delays in finding and approval. The deputy SPD has filled in as needed, but the lack of official approval can hamper effectiveness, especially in dealing with budgetary matters. The IPO has had only two SPDs, and the appointment process for both took on average a year to complete.

- Other key slots within the IPO, especially the deputy SPD, also have not been filled in a timely manner. The process by which individuals are selected and approved is cumbersome. Lack of key leaders has definitely hindered IPO effectiveness.
- The manning with DoD personnel slots at the IPO probably has presented more problems for the IPO management than for their civilian counterparts. Normal military turnover rates cause a continuous loss of corporate knowledge, and as the service with the largest military representation, Air Force personnel system rates make succession planning uncertain. DoD personnel who serve at NPOESS say they are taking a career risk, since they must move outside their normal manning structures. In some cases, junior officers have separated while at the IPO and have taken civilian positions in the IPO and elsewhere.
- The IPO personnel have noted that they are overworked at specific times, yet in general most agree that keeping the office small is advantageous to operations effectiveness. IPO managers have stressed the need to maintain flexibility in manning structure to deal with a variety of tasks. Also, the IPO must adjust to surges of activity during the year and over consecutive years, so a well-organized, efficient, and smaller IPO is actually preferred.

Conclusions

NPOESS has successfully managed to converge the planning and acquisition functions for the POES and DMSP follow-on programs as well as DMSP operations, but the convergence in POES operations and in other areas remain to be achieved. Essentially, a Presidential order gave the needed impetus for leaders to do what they had considered many times before, but had not been able to implement: merge the U.S. military and civilian weather satellite programs despite differences of culture, mission, and requirements, largely to increase efficiencies. This convergence shows that the differences can be overcome, and that certain program actions and strategies are very important to success.

This last section will sum up what RAND believes are the most valuable lessons learned, which may be applicable to other joint programs:

- Maximize preplanning as much as possible and include all key parties in that planning.
- The program should have highest possible signature support with each agency involved.

- Make sure the program leadership has a seat at the table of all forums or committees that have a significant impact upon the program. This is particularly true in budgetary matters.
- Maintain broad support for the program up the chain, continually educating key leaders on why the program exists and is required. Avoid reliance upon a single champion, instead creating a network of support among all important players.
- The early key documents, such as the primary MOA, should be as detailed as possible, specifically laying out roles and responsibilities. Leaving important details to a later date can seriously hamper a program, especially in budgetary matters.
- Early requirements documents must also be specific. Take whatever time is required so that all parties agree upon requirements before the program goes ahead. Also, ensure that requirements committees are staffed by all agencies and key mission users.
- All important requirements must face extensive budgetary audits, so program needs must be as detailed and quantifiable as possible.
- Spend significant effort upon maintaining consistent and well-defined budgets. As much as possible, continually educate comptrollers on program details.
- Physically collocate joint program personnel, so they become a team with a common purpose. Ensure that all member agencies have a common sense of joint ownership.
- The System Program Director is a pivotal position and should be politically and professionally acceptable to all agencies. A vacancy in this position can be detrimental, so place priority on a quick and efficient selection process.

Updates in 2001

As this document was about to go to print, the NPOESS IPO suggested an update to the RAND analysis, which is provided below.

As a joint agency program with no single overarching authority position, the NPOESS IPO depends very heavily on the Executive Committee (EXCOM) for decisionmaking and top level support. As described, the EXCOM is composed of the NOAA Administrator (Under Secretary of Commerce for Oceans and Atmosphere), NASA Deputy Administrator, and Under Secretary of Defense for Acquisition, Technology, and Logistics (USD(AT&L)). These individuals are at the policy making levels in their respective organizations and are able to give both guidance and binding direction. Additionally, the EXCOM serves as the

Acquisition Board and the DoD representative is also the Milestone Decision Authority within DoD. During the last year, the USD(AT&L) issued direction that day-to-day guidance of programs would be handled by the appropriate Assistant Secretary of Defense. However, USD(AT&L) did not transfer decisionmaking authority to the Assistant Secretary level because that is statutory. The result has been a significant slowdown in the IPO's ability to receive binding direction. Interviewees believe that the decision was probably meant to encompass only DoD programs, not joint agency programs covered by a PD and Secretary-level MOA, but the staff interpretation was all programs. The important lesson here for other joint agency programs is that the rules on assigning responsibility are far more important in cross-agency programs and should be carefully considered before any changes are made and should be, preferably, made part of the implementing agreements.

The NPOESS IPO was consciously established as a small program with a very flat management structure. The founders assumed that the program would manage their personnel needs within the allotted billets. Due to manning shortages by both career field and rank within the Air Force, the Air Force is unable to meet their committed personnel strength. If the NPOESS IPO were at a normal product center, this shortfall could be ameliorated by the local commander assigning additional resources (albeit at the expense of another program). Since the IPO is isolated, this cannot easily be done and the result has been an increasing shortage of Air Force officers in acquisition billets, the very billets the USAF was so insistent on protecting in the beginning. The lesson learned is that personnel manning levels should be explicitly agreed to in the beginning so that changing conditions do not automatically impact the office.

The budget cycles of DoC/NOAA, NASA, and DoD are different. The internal corporate processes have the basic overall structure of budget builds followed by submissions, reviews, rebuttals, etc,, but they move on different time cycles and even with different content. Whereas DoD has a service submission to OSD (POM), followed by the OSD approved budget that is jointly reviewed with OMB (BES), the DOC has a series of turnovers to higher levels with little collaborative review. Both processes work—the problems come when they are out of phase and one cycle ends up driving the other. The only common point that the IPO has found to resolve this issue is at OMB, a role never considered in the MOA. The only other alternatives are the EXCOM, but it is hampered by the previously mentioned DoD decision and the fact that the members are not within the comptroller chain, or OSTP, which has oversight responsibility for NPOESS but no budgetary authority. Use of OMB would be a reasonable solution except for differing agency rules on dealing with OMB. Again, the lesson learned is that different budget cycles produce severe friction that should be considered in establishing a joint agency program—raising the question of which agency has final budget reconciliation authority.

Appendix B. Arsenal Ship Case Study

Overview

For twenty months in 1996 and 1997, the U.S. Navy and DARPA operated a joint program office to manage the acquisition of a new type of fleet vessel called the Arsenal Ship. For those pursuing similar cooperative projects among two or more government agencies, the Arsenal Ship story can serve as a valuable case study, for it represented a new operational methodology and emphasized a streamlined and non-traditional management structure and process. In the end, the Arsenal Ship project was cancelled, but interesting lessons resulted which provide insight into how the program operated, how the two agencies cooperated, and reasons why the effort eventually failed.

Sponsored by DARPA, RAND's National Defense Research Institute (NDRI) conducted a definitive analysis of the Arsenal Ship program and in 1999 published a document titled *The Arsenal Ship: Acquisition Process Experience, Contrasting and Common Impressions from the Contractor Teams and Joint Program Office.*[97] While other sources were used in preparing the case study, the RAND study represents the primary source of information and anyone seeking further details about the Arsenal Ship program should read this document. We will cover the general details of the Arsenal Ship program in this summary to set the stage for lessons learned. However, for this case study, we are most interested in issues pertaining to the formation and operation of the JPO, so only these will be covered in more detail later.

Program Goals and Structure

The idea and planning for a new class of naval expeditionary force warfighting vessel, which came to be called the Arsenal Ship, took shape in 1995. The champion of the program was the Chief of Naval Operations, Admiral Jerry M. Boorda, who looked for a new type of ship to meet existing needs at much lower costs. Planners looked to satisfy requirements in regional naval conflicts by providing commanders with massive firepower, long range strike, flexible

---

[97]Robert S. Leonard, Jeffrey A. Drezner, and Geoffrey Sommer, *The Arsenal Ship: Acquisition Process Experience, Contrasting and Common Impressions from the Contractor Teams and Joint Program Office*, MR-1030-DARPA (Santa Monica, California: RAND, 1999).

targeting, and even theater defense with a low visibility ship carrying hundreds of vertical launch system (VLS) cells. The key elements of the design included:

- Provide around 500 VLS cells, which could launch Navy and joint munitions in support of land campaigns.
- Ensure the possibility of off-board control by integrating a combat system called Cooperative Engagement Capability.
- Allow for the possibility of incorporating ship design features like survivability and self-defense at a later date.
- Emphasize low ownership cost by using innovative operational and maintenance methodologies.
- Further emphasize cost savings by designing for a small crew size not to exceed 50 personnel.
- View CAIV and employ only off-the-shelf systems and technology.

The Arsenal Ship program manager stated, "In the face of limited budget levels, the use of acquisition reform initiatives and streamlined contracting methods were paramount to meet the basic requirements of the Arsenal Ship in an affordable manner."[98] Contractors and the government would work very closely to quickly create a demonstration vessel for a fixed Unit Sail-away Price (USP) of $550M (FY96). It would be evaluated to see how well it integrated into fleet operations and in support of ground forces. No other ship had been so specifically designed to support primarily land operations with massive offshore firepower. Both the Air Force and aircraft carrier attack wings had been filling that role to date. Not since the creation of ballistic submarines in the 1960s had the Navy looked at a new class of fleet vessel. The Arsenal Ship was not only a new phenomenon, but also a big risk for a tradition-minded Navy.

The program officially began on March 18, 1996 when the Assistant Secretary of the Navy for Research Development and Acquisition and the Director of DARPA signed a Joint Memorandum, which established the Arsenal Ship Program. This provided the Director of DARPA, Commander, NAVSEA, and Chief of Naval Research "with precepts regarding the basic requirements, goals, and acquisition strategy for the program."[99] The Memorandum delineated few details, but rather gave the sufficient high-level efficacy to aggressively move the Arsenal Ship program forward. Two months later, in May 1996, Navy and DARPA signed the MOA with the purpose to "establish a joint Navy/DARPA agreement as to the objectives, roles, and responsibilities, schedule, and funding for the

---

[98]Charles S. Hamilton, Capt USN, Arsenal Ship Program Manager, *DARPA—Arsenal Ship Lessons Learned,* 31 December 1997.

[99]Leonard, et. al., op. cit., p. 133.

Arsenal Ship demonstration program."[100] With significantly more details, the MOA described the background, technical objectives, and financial goals. Also, it officially created the Arsenal Ship Joint Program Office (ASJPO), structured outside of normal DOD acquisition rules, utilizing special DARPA acquisition rules.

As the first two Memoranda initiated the program and provided official stamps of approval, two other documents were created to give contractors more substance on what the ASJPO really intended.

The Ship Capabilities Document (SCD), as the name implies, defined essential technical attributes and expected capabilities of the Arsenal Ship. The CONOPS document, produced by the Navy, concentrated on the bigger picture, focusing on how the vessel was expected to operate in the fleet. Neither of these key documents is very long (less than 10 pages each), compared to traditional programs that required detailed tomes of technical specifications. This emphasis on less detail and more trust in contractors to meet general requirements was a key element of the Arsenal Ship program. We will focus upon these early program documents below to gauge what worked and what may have been missing.

From the onset, the key goals of the Arsenal Ship program were:

1. Demonstrate affordable acquisition of a capable product,
2. Significantly increase utilization of commercial practices and technology, and most importantly for this study,
3. Demonstrate that reformed acquisition processes can work.

The Arsenal Ship program had six distinct phases, of which only the first two were completed before cancellation in October 1997.

- Phase I: Lasted six months and included cost-performance studies focussing on initial design concepts. Five contracting teams made bids.
- Phase II: 12 months long, where three surviving teams evolved initial concepts toward functional designs.
- Phase III: This phase would have lasted 33 months, where a single winning contractor was to design and construct an Arsenal Ship Demonstrator.
- Phase IV: 12 months long, was to be the test and evaluation period of the Demonstrator against the SCD and CONOPS.

---

[100]Ibid., p.137.

- Phase V: Duration to be determined, where the contractor would build up to a total of six operational vessels, including upgrading the Demonstrator to operational status.
- Phase VI: Also undetermined, was to see if service life-cycle support tasks would be required for the operational vessels.

From the MOA, program management fell jointly to the Navy and DARPA, while the Program Manager (PM) reported to DARPA. The PM developed the overall program plan and also reported directly to two external committees for guidance:

- The Steering Committee, which approved the initial program plan and was thereafter expected to conduct quarterly reviews to check on status and provide help as needed. This committee consisted of:
    - Director, Tactical Technology Office (TTO), DARPA (chair)
    - Deputy Assistant Secretary of the Navy (DASN, Ships)
    - Assistant Director of DARPA, TTO for Maritime Programs
    - Director, Surface Warfare Plans/Programs/Requirements Branch
    - PEO for Surface Combatants
    - Office of Naval Research
- The Executive Committee was to review the program at major milestones and evaluate program cost thresholds. It consisted of the following members:
    - Director, DARPA (chair)
    - Assistant Secretary of the Navy
    - Director of Surface Warfare
    - Commander, NAVSEA
    - Chief of Naval Research

Program Acquisition Strategy

The initial industry funding plan and cost schedule is shown below in Figure B.1.

The acquisition strategy was a complete departure from traditional Navy shipbuilding programs. The Arsenal Ship project was a non-ACAT (non-Acquisition Category) entity that did not conform to normal MDAP policies. It was never subject to either the DOD 5000.1 or 5000.2-R acquisition regulations. The key elements to the process included:

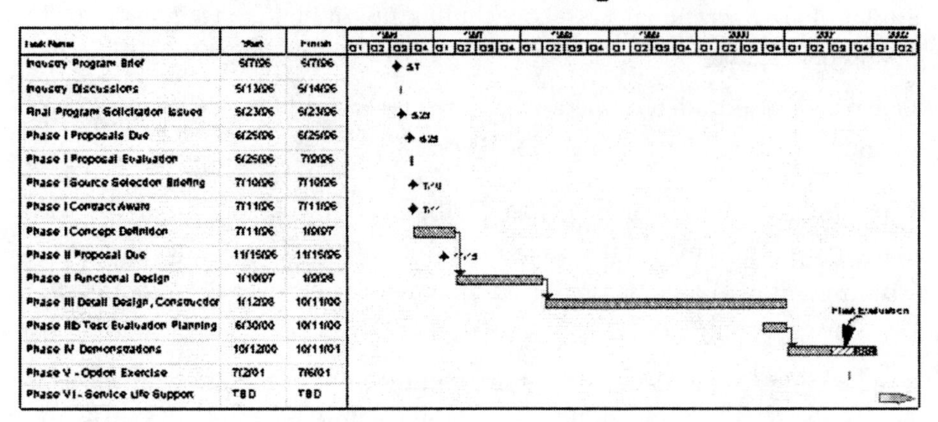

Fiscal Year Funding Profile

Phase I (Multiple Awards) $1M ea
Phase II (Two Awards) $15M ea
Phase III } **(One Award)** $167M $119M $103M
Phase IV } $25M*

* **Phase IV T&E distribution between Gov't and Contractor to be determined**

Source: DARPA-Arsenal Ship Lessons Learned-31 Dec 97

Figure B.1: Initial Schedule and Funding Profile

- Relatively few broad performance goals were used to define desired system capabilities.
- Competing contractor teams were given full design responsibility. Also, the program excluded regulations on Government Furnished Equipment (GFE).
- The ASJPO was a very small group, never exceeding 20 personnel. Compared to traditional JPOs, this was a significant reduction.
- The only firm requirement was affordability, while there was also strong emphasis on small crew size, since large crews meant large costs.
- The program utilized Integrated Product and Process Teams. These were government, industry and ASJPO teams that worked together to achieve program goals in a less adversarial manner than traditional programs. Also, these teams designed the ship as a whole, as opposed to traditional methods of designing subsystems and the hull separately.

- The program was structured around DARPA's Section 845 Other Transaction Authority (OTA).[101] This allowed the ASJPO to operate outside traditional DoD acquisition regulation, concentrating on streamlined acquisition and innovative practices.

As the RAND study indicates, "The absence of detailed requirements and system specifications, coupled with the transfer of design responsibility from the government to the contractor teams, was the most striking aspect of the acquisition approach."[102] This came to be unofficially called "common sense acquisition."

For such a radical departure of methodology to succeed, the ASJPO had to also operate in new ways. With very few personnel, they could not and were not expected to be involved with every detail in program management. The small size was mostly successful because each individual was handpicked, highly motivated, and specifically qualified for the program. Key experts were borrowed from other agencies at critical junctures, so that the small ASJPO could leverage capabilities far beyond their office. While there were problems with both the acquisition approach and the use of a small JPO, both of these aspects turned out to be clearly positive in the end.

The acquisition strategy was certainly not perfect, for in the end it became one of the reasons for the program's failure. We will list some of the bigger problems here and expand upon those directly related to the JPO later on.

- The schedule and cost structure for the Arsenal Ship was very ambitious. Few problems appeared in Phase I, but the Phase II contractor teams all believed that funding for Phase III was far too low. This was due to serious disagreements about required resources, development tasks, and overoptimistic initial designs.
- Since the ASJPO operated outside traditional bounds, they had no authority to compel traditional support agencies for help. The Naval Surface Warfare Center, other federal labs, and much of the naval bureaucracy were not entirely cooperative. All agree that had the program continued past Phase II, when such external resources would have been critical, this problem would only have gotten much worse.
- Production costs of the Arsenal ships were initially fixed at the USP. The fixed maximum price became a sticking point in Phase II as the teams

[101] Section 845 of the National Defense Authorization Act of 1994 authorized DARPA to conduct prototype projects of weapons systems under the authority of 10 U.S.C 2371. Section 804 of the 1997 National Defense Authorization Act extended this authority to military departments and other DoD components until 1999.

[102] Leonard, et. al., op. cit., Summary, p. xv.

complained that an irrevocable USP was impractical, due to things learned in the earlier Phase. Contractors believed that they would end up operating in a fixed-priced development scheme, while the ASJPO disagreed.

Maintaining Stakeholder Support

The official reason given for the failure of the Arsenal Ship program was that there were not sufficient funds due to other priorities. There were other reasons, but most of all, the course of Arsenal Ship was set early by insufficient Navy support and critical lack of funding for Phase III. Had the Navy bureaucracy and congressional players really wanted to save the program and correct the early flaws, Arsenal Ship may have survived to production. While the program had a powerful champion, a capable and motivated JPO, and sufficient contractor support, the lack of general acceptance within the greater naval leadership community doomed the program. The very nature of the acquisition strategy, while proving dynamic and effective, alienated traditional players and created support for competing programs. The contractors could see problems rising quickly within the Navy and Congress during Phase II, so their commitment rapidly waned, assuring that nobody would risk further development.

Since the Arsenal Ship program was cancelled early, it is difficult to ascertain the general success or failure of the ASJPO and the acquisition strategy. The RAND study makes three conclusions in regard to the original program goals.

1. The program did NOT develop enough to demonstrate the capability of the weapon system.
2. The program did mature to the point where the RAND authors could say it successfully leveraged commercial practices and technologies.
3. A comprehensive conclusion of the reformed acquisition process could not be made. While the new process mostly worked well for the early Phases, only two of six were completed and thus a full evaluation whether that success would have continued could not be made. The relationship between government and the final winning contractor would have no doubt been different in the later phases when the lone contractor no longer had to worry about competition.

Nevertheless, many valuable lessons from the Arsenal Ship program are directly attributable to other federal joint programs, specifically in regard to the third goal to see how the ASJPO operated and how the Navy initiated and sponsored the program from inception to its eventual cancellation.

Lessons Learned For Future Interagency Acquisition

This section will focus upon the lessons learned from the Arsenal Ship program that were documented in the RAND report, as well as from a final report from the Program Manager. The conclusions and observations were made after extensive discussions with the ASJPO and contractor teams. As with any large-scale program, the merits of the actual project often become a secondary issue compared to the political issues and budget battles. The original idea was good enough and probably deserved further implementation. The Arsenal Ship did not fail for the idea, but for all those other reasons, which are just as important.

Need to Maintain Support and Funding

The key reason for the program's immediate failure in 1997 was lack of Navy and congressional support, which led to insufficient funding of Phase III. From the beginning, the program suffered from an ongoing tale of misunderstandings over monies and a desire for the Arsenal Ship supporters to keep the program moving forward despite serious financial problems.

To pay for Phases I-IV, NAVSEA originally estimated $600-700 million, using traditional acquisition pricing methods. This estimate did not include the Phase III construction of an *all-new* vessel, but focused primarily upon nonrecurring engineering tasks. The estimate was cut in half simply because the program utilized DARPA's streamlined acquisition process and emphasized CAIV. So, the ASJPO and Phase III winning contractor would be allocated roughly $350 million and an existing Navy destroyer for conversion to an Arsenal ship. By the end of Phase I, all parties involved knew that Phase III was significantly underfunded, but through a series of misunderstandings and oversights, the issues did not get pressed.

The contractors in Phase II did not press the underfunding issues because they wanted to first win the Phase III competition then deal with the issues from a stronger position. The three winning contractors clearly perceived a funding problem as they moved into Phase II, but they had several motives to continue. Each wanted to gain a competitive advantage for future business opportunities. If they stayed in the Arsenal Ship program, the funding problems could have been fixed, but if not, they could leverage their experience to compete for future programs that competed with Arsenal Ship, specifically the SC/DD 21 destroyer.

More importantly, the ASJPO also admitted afterward that they knew the program was not properly funded. The JPO team was committed to completing the program of record in the specified time for the original amount of money. Since they had weak support in the broader Navy and Congress, the ASJPO feared that any request for increased funds would quickly give their opponents the means to cancel the program. They hoped to push forward into Phase III far enough to either gain more support or prove the program concept and then gain necessary funding to complete the phase.

While many mistakes were made at all levels in regards to the underfunding, the key fact is that the program never really had a solid financial base upon which to stand. The top leaders had misunderstandings on issues like whether an entirely new ship would be built or whether an existing destroyer would be converted. In a program that suffered as much controversy as the Arsenal Ship from inception, it is clear that the fundamental issues still could have been more clearly agreed upon in the foundation documents.

Need to Maintain Internal and External Support

The RAND study states,

> Absent explicit support by the chief executive or an organized congressional lobby, congressional support for any acquisition program is shaped by the procuring service's resolve to move the program forward. When a user service is behind a particular program, it will give that program the time and energy, at all levels, to make it as successful as possible in the eyes of Congress and the public. The service undertakes a parallel effort to convince Congress of how essential the program is to national security. The service educates and shapes the thought processes of congressional advisors as well, using a combination of hard work and skilled marketeering.[103]

The Arsenal Ship program never received this essential support from the general Navy community, either because they simply did not believe in the CONOPs, or the program threatened existing programs. If the Navy community in general does not support one of their own programs, it must be obvious that congressional support will also not materialize. There are four specific reasons why the program began despite lack of broad support, and why those reasons were not enough to get beyond Phase II.

- The Chief of Naval Operations in 1995 was Admiral Boorda. The Arsenal Ship program was his and he alone provided the clout necessary to create not only a new type of naval vessel, but also to obtain it using non-traditional means of acquisition. In May 1996, the Admiral died unexpectedly, only two months after the first Memorandum. The Arsenal Ship had lost its powerful champion. Without Admiral Boorda, no top leader remained to advocate Arsenal Ship among powerful Navy and congressional groups. The program was so dependent upon him that without his continued presence, failure was in reality only a matter of time.

---

[103]Ibid., p. 83.

- In large measure, the lack of support from the Navy did not arise because of the idea. The Arsenal Ship was deemed a very promising weapon system by many in the Navy. However, its intended mission was currently covered first by aircraft carrier battle groups and secondly by Air Force long-range precision bomber forces. A successful Arsenal Ship program could have undermined the future procurements of traditional carriers and B-2 bombers, programs with very powerful lobbies within the Pentagon.
- The traditional Navy acquisition corps was not happy with the Arsenal Ship program, since the program was specifically designed to circumvent them in the desire to achieve superior results. A successful demonstration of streamlined acquisition could easily undermine the need for long-standing and large numbers of procurement offices. The clearest manifestation of this was the general lack of support that the naval weapons labs provided for the Arsenal Ship contractors and ASJPO.
- Even if some liked the Arsenal Ship idea, in the congressional budget battles, the new ship could take money away from other programs. The Arsenal Ship had a near competitor in the more traditional SC/DD 21, a product of the five-year Defense Plan. If all six Arsenal ships had been built, they were expected to cost $3-4 billion, spread over 12 years. The Navy saw no need to fund both programs, and quickly incorporated out year funds of the Arsenal Ship into the SC/DD 21 program. The new destroyer also took on several design and operational aspects of the Arsenal Ship. The SC/DD 21 would never be as useful in the specific mission envisioned for the Arsenal Ship, but the destroyer was certainly more versatile inside overall Navy doctrine. The choice between Arsenal Ship and SC/DD 21 became either/or, and for the Navy and the acquisition corps, the choice was very simple.
- Admiral Boorda's brainchild also threatened many agendas within Congress. In its very short existence, the Arsenal Ship program proved a volatile political issue, threatening existing allocations of monies and future influence. While most weapons systems are simply newer versions of old ones, they tend to threaten no special interests and elicit no real changes in force structure. However, Arsenal Ship could have dramatically affected whether traditional ships were built, whether large numbers of manpower would be required, and also whether manufacturing and support jobs would be lost. Three specific groups with congressional lobbies were seriously threatened by Arsenal Ship.
  - Manufacturers of competing weapon systems, such as carriers and B-2 bombers.

- Host facilities of competing weapon systems. Large numbers of military and civilian jobs could be displaced if competing systems lost support in favor of the Arsenal Ship.
- The large, traditional Navy acquisition infrastructure. Again, large numbers of jobs were at stake.

Lobbies representing Newport News Shipbuilding and Northrop Grumman pursued the Arsenal Ship in Congress, ensuring that their very expensive programs would remain.

Maintaining Programmatic Innovation While Acknowledging Existing Requirements Processes

As stated above, the small ASJPO and streamlined acquisition process could in general be called successful, but not without a downside. The very nature of the Arsenal Ship procurement process created reasons for suspicion within not only DoD, but also Congress. Normal oversight expected in the MDAP process was absent and there were none of the normal milestones, documents or requirements of traditional acquisition methods. Enemies in Congress used this as a means to attack the Arsenal Ship program, sighting lack of oversight and accountability. Right or wrong, the perception was enough.

Traditional ACAT I programs have specific sets of requirements that many have come to depend upon in any evaluation. Normally, the program office issues a MNS, which outlines why the new system is needed and how it will be utilized in the fleet. Other documents follow, including the ORD and the Analysis of Alternatives (AOA), as well as detailed design reviews within the program milestones. The ASJPO effectively covered most of the bases with the Memorandums, the SCD, and the CONOPs, but in one sense, they failed to heed enough tradition by completely ignoring the MNS. The RAND study states,

> Had Adm. Boorda issued a rudimentary Mission Need Statement to underpin the program's CONOPS and SCD, the Navy would have found it difficult to waver in its support of the Arsenal Ship concept in his absence. The program's acquisition approach is directly responsible for the absence of such a document. Those who preside over future streamlined programs should take note of the importance of having such a document to ensure that the military utility of the weapon system concept has been established.104

Admiral Boorda and the ASJPO may have been able to keep the program going by vision and willpower, but after his death, the lack of consensus that an MNS could have provided seriously hampered the program. It is vitally important that all key details and responsibilities of the program are clearly enunciated in

[104]Ibid., p. 84.

writing before real activity begins. The major players who have significant budgetary and political influence over the program must sign off on at least the basic idea, or the list of opponents will grow as leaders are forced to support their own priorities.

Insights and Implications

There are many more insights from the Arsenal Ship program that can be useful to future interagency acquisition and IPOs, whether they operate in a streamlined methodology or in traditional acquisition programs. If innovative strategies are implemented, similar to that described above, it is key that all the elements of such an approach go together, or not at all. One cannot have a small joint program office unless contractors receive significant design authority and the documents such as the SCD are kept short and relatively simple. Each element is mutually enabling and reinforcing of the other elements.

If future innovative programs can use the streamlined approach, the following provisions will aid implementation:

- The program office must be staffed with high-quality and motivated personnel, both on the government and industry side. These people must believe in the innovative process and expect proper rewards or promotions when they work in the program office.
- Acquisition process change must be a top priority, so that innovative methods find a receptive environment.
- Key program individuals require significant flexibility, so they can make timely decisions and take advantage of opportunities as they arise.
- Contractors must be made responsible for outcomes, so they feel closely connected to program success.
- A program must have stable and consistent government commitment at all levels of a plan that has been accepted by all relevant stakeholders.

This case study has been presented to elicit further thoughts on how to organize a future IPO and describe some pitfalls that could hinder success of a new program. The Arsenal Ship example has many unique qualities, but it can reinforce a few vital issues that any JPO or IPO should consider.

- Examine the documents that create and organize the joint program. Do they lay out specific requirements and responsibilities, and also cover the necessary political and bureaucratic bases? The ASJPO never achieved consensus in the naval community for the need for their weapon system. Without Navy support, they never achieved congressional support, and finally lost funding. It is possible that a simple MNS early in the program could have saved the day later on.

- Identify single point failures that could derail program success. For the Arsenal Ship, Admiral Boorda became a single point failure, for when he died in early 1996, the program lost its champion. Without him, the ASJPO could not deflect opponents from killing the program.
- Review rival programs that could outbid yours on IPLs. If a program is successful, most often it will displace something else in today's era of zero-sum game budgets. A successful Arsenal Ship threatened a large host of powerful interests, in the Navy and Congress.
- Identify and assess what external agents are needed to ensure program success. The Arsenal Ship JPO had noted trouble with all external agencies, because they alienated nearly everyone in the traditional community by using a new and controversial acquisition approach. Even with an innovative and creative acquisition strategy that was generally deemed a success, the ASJPO found out they could not operate in isolation from, or at the expense of, other programs.
- Identify funding schemes to ensure the program can actually be funded at all critical milestones. Nearly all the Arsenal Ship players knew that the program was critically underfunded by early Phase II. Even if the program had continued, the requirement for more funds alone could have meant outright cancellation.

All these points should be nothing new to experienced acquisition managers, but on the other hand, these are lessons learned from a recent JPO operated by "highly qualified," experienced personnel. In hindsight, there were areas that could have been handled differently. Insights such as these are of interest to any future IPO or JPO.

Bibliography

Baca, Jim, Deputy Technical Director, "ACTD: AFOTEC Perspective," briefing, c. January 2000.

Birkholz, Eric L. Birkholz, Editor, Erica J. Claflin, Sr. Associate Editor, and Nora O'Hagan, Assistant Editor, *Congressional Yellow Book,* Volume 26, Number 2. Washington, D.C.: Leadership Directories, Inc., Summer 2000.

Birkler, John, C. Richard Neu, and Glenn A. Kent, *Gaining New Military Capability: An Experiment in Concept Development*, Santa Monica, California: RAND MR-912-OSD, 1998.

Birkler, John, Giles Smith, Glenn A. Kent, and Robert V. Johnson, *An Acquisition Strategy, Process, and Organization for Innovative Systems*, Santa Monica, California: RAND MR-1098-OSD, 2000.

Bracken, Paul, *Strategic Planning for National Security: Lessons from Business Experience,* Santa Monica, California: RAND N-3005-DAG/USDP, February 1990.

Bracken, Paul, John Birkler, and Anna Slomovic, *Shaping and Integrating the Next Military: Organization Options for Defense Acquisition and Technology,* Santa Monica, California: RAND DB-117-OSD, 1996.

Bradbury, Lt Col Chuck, "Joint Experimentation Program CINC's ACTD Representatives," USFORSCOM briefing, 16 May 2000.

Cahan, Bruce, President, Urban Logic, Inc., *Financing the NSDI: Aligning Federal and Non-Federal Investments in Spatial Data, Decision Support and Information Resources, Executive Summary,* draft report for public comment, February 29, 2000, found at http://www.fgdc.gov/funding/urbanlogic_exsum.pdf.

Carlson, Bruce, Lieutenantt General, USAF, Director J-8, "ACTDs: 'J-8 Perspective,'" briefing, 11 January 2000.

Catington, Lieutenant Colonel Richard C., USAF, Lieutenant Colonel Ole A. Knudson, USA, and Commander Joseph B. Yodzis, USN, *Transatlantic Armaments Cooperation: Report of the Military Research Fellows DSMC 1999-2000,* Fort Belvoir, Virginia: Defense Systems Management College Press, August 2000.

Cochrane, C. B., and G. J. Hagan, *Introduction to Defense Acquisition Management,* 4th ed., Fort Belvoir, Virginia: Defense Systems Management College Press, June 1999.

Cohen, William S., Secretary of Defense, and General Henry H. Shelton, Chairman of the Joint Chiefs of Staff, *Report to Congress: Kosovo/Operation Allied Force After-Action Report*, Washington, D.C.: Department of Defense, 31 January 2000.

Covault, Craig, "China Seen As Growing Reconnaissance Challenge," *Aviation Week and Space Technology*, August 7, 2000, pp. 65-66.

________, "NIMA Infotech Retools U.S. Space Recon Ops," *Aviation Week and Space Technology*, August 7, 2000, pp. 62-65.

Crawley, Vince, "Air Force Wants To Charge Other Services To Use Its Satellites," *Military Space*, Volume 17, No. 2, January 18, 2000.

________, "Space Chief Wants Affordable Discoverer II," *Military Space*, Volume 17, No. 2, January 18, 2000.

Defense Systems Management College, *Acquisition Strategy Guide,* 4th ed., Fort Belvoir, Virginia: Defense Systems Management College Press, December 1999.

Department of the Air Force, *America's Air Force Vision 2020, Global Vigilance, Reach, and Power*, Washington, D.C.: June 2000.

————, *Global Engagement: A Vision for the 21st Century.* Headquarters, United States Air Force, Washington, D.C.: 1997.

________, *The Aerospace Force: Defending America in the 21st Century; A White Paper on Aerospace Integration*, Washington, D.C.: 2000.

Department of Defense, DoD Directive (DODD) 5000.1, *Defense Acquisition*, Washington, D.C.

________, DoD Regulation 5000.2-R, *Mandatory Procedures for Major Defense Acquisition Programs (MDAPs) and Major Automated Information System (MAIS) Acquisition Programs*, Washington, D.C.

Department of the Interior, *Charter, National Satellite Land Remote Sensing Data Archive Advisory Committee*, 30 September 1999, available at http://edc.usgs.gov/programs/nslrsda/advisory/charter7.htm.

———, *Justification, National Satellite Land Remote Sensing Data Archive Advisory Committee*, 30 September 1999, available at http://edc.usgs.gov/programs/nslrsda/advisory/justification.html.

———, National Satellite Land Remote Sensing Data Archive Advisory Committee, Memorandum to the Secretary of the Interior, re: *National Satellite Land Remote Sensing Data Archive Policy White Paper*, January 25, 1999, found at http://edc.usgs.gov/programs/nslrsda/advisory/whitepaper.html.

Driesman, Andrew S., *Discoverer II Objective System Bus Concept Development Summary*, white paper, Baltimore, Maryland: Johns Hopkins University Applied Physics Laboratory, 10 December 1998.

Eash, Joseph J., III, "Harnessing Technology for Coalition Warfare," *NATO Review*, Summer/Autumn 2000, pp. 32-34.

Eller, Lieutenant Colonel Barry A., *Joint Program Management Handbook*, 2nd ed., Fort Belvoir, Virginia: Defense Systems Management College Press, July 1996.

EROS Data Center, *Earthshots*, Sioux Falls, South Dakota, found at http://edc.usgs.gov/earthshots/slow/SiouxFalls/SiouxFalls.

Executive Order 12906, *Coordinating Geographic Data Acquisition and Access: The National Spatial Data Infrastructure*, published in the April 13, 1994, edition of the Federal Register, Volume 59, No. 71, pp. 17671-17674.

Federal Geographic Data Committee, *Report of the FGDC Design Study Team on Financing the NSDI*, March 14, 2000.

Friedman, Bruce A., MD, "Radiology Management," November/December 1997.

Gansler, Jacques S., Under Secretary of Defense (Acquisition and Technology), *Keynote Address: Advanced Program Management Course 99-02*, Defense Systems Management College, Fort Belvoir, Virginia, May 10, 1999.

———, *The Revolution in Military and Business Affairs: The Road Ahead*, remarks, Defense Systems Management College, Fort Belvoir, Virginia, May 8, 2000.

General Accounting Office, *Defense Acquisitions: Improvements Needed in Military Space Systems' Planning and Education*, GAO/NSIAD-00-81, Washington, D.C.: United States General Accounting Office, May 2000.

________, *Military Personnel: Systematic Analyses Needed to Monitor Retention in Key Careers and Occupations*, GAO/NSIAD-00-60, Washington, D.C.: United States General Accounting Office, March 2000.

Grant, Jeff, et. al., *Ensuring Successful Implementation of Commercial Off-the-Shelf Products (COTS) in Air Force Systems.* Study sponsored by the Air Force Scientific Advisory Board and conducted in 1999. Washington, D.C., February 2000.

Hall, R. Cargill, NRO Historian, "The NRO at Forty: Ensuring Global Information Supremacy," unpublished article, c. Spring 2000.

Hough, Michael, Major General, USMC, Director, Joint Strike Fighter Program Office, "The Affordable Solution—JSF," briefing, found at http://www.jast.mil/.

Hughes, Mark, Colonel, USAF, Program Director, "DARPA Tech Discoverer II," briefing, June 1999.

________, "Space-Based HRR-GMTI/SAR Demonstration Program," briefing, 5 November 1998.

Joint Chiefs of Staff, *Joint Intelligence Support to Military Operations*, Joint Pub 2-01, Washington, D.C., 20 November 1996.

________, *National Intelligence Support to Joint Operations*, Joint Pub 2-02, 28 September 1998.

Joint Strike Fighter Program Office, *Joint Strike Fighter White Paper*, Arlington, Virginia: 2 December 1999, found at http://www.jast.mil/html/jsfwhitepaper.htm.

Jones, Wilbur D., Jr., *Congressional Involvement and Relations: A Guide for Department of Defense Acquisition Managers*, 4th ed., Fort Belvoir, Virginia: Defense Systems Management College Press, April 1996.

Leonard, Robert S., Jeffrey A. Drezner, and Geoffrey Sommer, *The Arsenal Ship Acquisition Process Experience: Contrasting and Common Impressions from the Contractor Teams and Joint Program Office*, Santa Monica, California: RAND MR-1030-DARPA, 1999.

Lorell, Mark, Michael Kennedy, Julia Lowell, and Hugh Levaux, *Cheaper, Faster, Better? Commercial Approaches to Weapons Acquisition*, Santa Monica, California: RAND MR-1147-AF, 2000.

National Imagery and Mapping Agency, *NIMA Strategic Plan: Guaranteeing the Information Edge*, Washington, D.C.: 1999.

Nefzger, David F., *Collection Management: Theory, Process, and Practice*, Washington, D.C.: Defense Intelligence Agency, February 1999.

O'Connor, Michael J., "Advanced Concept Technology Demonstration: ACTD Process Overview," briefing, c. January 2000.

Office of the Secretary of the Air Force/Assistant Secretary for Acquisition, National Reconnaissance Office, and Defense Advanced Research Projects Agency, *Memorandum of Agreement: Spacebased Radar Risk Reduction and Demonstration Program*, Washington, D.C.: February 1998.

Office of Technology Assessment, Congress of the United States, *Assessing the Potential for Civil-Military Integration: Technologies, Processes, and Practices*, Washington, D.C.

Pace, Scott, Brant Sponberg, Molly Macauley, *Data Policy Issues and Barriers to Using Commercial Resources for Mission to Planet Earth*, Santa Monica, California: RAND DB-247-NASA/OSTP, 1999.

Pace, Scott, Gerald Frost, Irving Lachow, David Frelinger, Donna Fossum, Donald K. Wassem, Monica Pinto, *The Global Positioning System: Assessing National Policies*, Santa Monica, California: RAND MR-614-OSTP, 1995.

Poussard, Ron, Contracting Officer, "Discoverer II: Contracting to Meet Program Objectives," briefing, 5 November 1998.

Rinaldi, Steven M., *Sharing the Knowledge: Government-Private Sector Partnerships to Enhance Information Security*, INSS Occasional Paper 33, Colorado Springs, Colorado: Institute for National Security Studies, U.S. Air Force Academy, May 2000.

Sarsfield, Liam, "The Application of Best Practices to Spacecraft Development: An Exploration of Success and Failure in Recent Missions," RAND briefing, Summer 2000.

The White House, Office of the Press Secretary, *Fact Sheet: U.S. Polar-Orbiting Operational Environmental Satellite Systems*, May 10, 1994.

Thirtle, Michael R., Robert V. Johnson, and John L. Birkler, *The Predator ACTD: A Case Study for Transition Planning to the Formal Acquisition Process*, Santa Monica, California: RAND MR-899-OSD, 1997.

United States Air Force Scientific Advisory Board, *Report on a Space Roadmap for the 21st Century Aerospace Force*, Volume 1: Summary, SAB-TR-98-01, Washington, D.C.: November 1998.

________, *Vision of Aerospace Command and Control For the 21st Century*, SAB-TR-96-02ES, Washington, D.C.: November 1996.

United States Congress, House of Representatives Permanent Select Committee on Intelligence, *Intelligence Authorization Act for Fiscal Year 2001*, Committee Report, Washington, D.C.: Spring 2000.

United States Congress, Senate Select Committee on Intelligence, *Authorizing Appropriations for Fiscal Year 2001 for the Intelligence Activities of the United States Government and the Central Intelligence Agency Retirement and Disability System and for Other Purposes*, Committee Report, Washington, D.C.: Spring 2000.

United States Geological Survey, *National Satellite Land Remote Sensing Data Archive*, 11 February 1998, found at http://edc.usgs.gov/programs/nslrsda/overview.html.

________, *USGS Web Site for Memorandums of Understand (MOU)*, found at http://www.usgs.gov/mou/.

________, *United States Geological Survey Manual*, Section 500.3, "Policy on Work for Other Federal Agencies," April 26, 1994.

________, *USGS National Mapping Program Activities with Foreign Governments and International Organizations*, 4 November 1998, found at http://mapping.usgs.gov/html/international/international.html.

________, *USGS Vision, Mission, Strategic Direction*, found at http://www.usgs.gov/bio/USGS/mission.html.

Wall, Robert, "U.S., U.K. Missile Needs Seen Converging," *Aviation Week and Space Technology*, August 7, 2000, p. 46.